MATLAB® Programming for Biomedical Engineers and Scientists

MATLAB® Programming for Biomedical Engineers and Scientists

Andrew P. King

Paul Aljabar

Academic Press is an imprint of Elsevier
125 London Wall, London EC2Y 5AS, United Kingdom
525 B Street, Suite 1800, San Diego, CA 92101-4495, United States
50 Hampshire Street, 5th Floor, Cambridge, MA 02139, United States
The Boulevard, Langford Lane, Kidlington, Oxford OX5 1GB, United Kingdom

Notices

Knowledge and best practice in this field are constantly changing. As new research and experience broaden our understanding, changes in research methods, professional practices, or medical treatment may become necessary.

Practitioners and researchers must always rely on their own experience and knowledge in evaluating and using any information, methods, compounds, or experiments described herein. In using such information or methods they should be mindful of their own safety and the safety of others, including parties for whom they have a professional responsibility.

To the fullest extent of the law, neither the Publisher nor the authors, contributors, or editors, assume any liability for any injury and/or damage to persons or property as a matter of products liability, negligence or otherwise, or from any use or operation of any methods, products, instructions, or ideas contained in the material herein.

Library of Congress Cataloging-in-Publication Data
A catalog record for this book is available from the Library of Congress

British Library Cataloguing-in-Publication Data
A catalogue record for this book is available from the British Library

ISBN: 978-0-12-812203-7

Publisher: Mara Conner
Acquisition Editor: Tim Pitts
Editorial Project Manager: Anna Valutkevich
Production Project Manager: Jason Mitchell
Designer: Greg Harris

Typeset by VTeX

Dedication

A.P.K. – For Elsa, Helina, and Nathan.

P.A. – For Jo, Owen, and Leila.

"When somebody has learned how to program a computer ... You're joining a group of people who can do incredible things. They can make the computer do anything they can imagine."

Sir Tim Berners-Lee

Contents

About the Authors

Andrew P. King has over 15 years of experience of teaching computing courses at university level. He is currently a Senior Lecturer in the Biomedical Engineering department at King's College London. With Paul Aljabar, he has designed and developed the Computer Programming module for Biomedical Engineering students upon which this book was based. Andrew and Paul have been delivering the module together since 2014. Between 2001 and 2005, Andrew worked as an Assistant Professor in the Computer Science department at Mekelle University in Ethiopia, and was responsible for curriculum development, and design and delivery of a number of computing modules. Andrew's research interests focus mainly on the use of medical images to tackle the problem of organ motion and to analyze motion to extract clinically useful biomarkers (Google Scholar: https://goo.gl/ZZGrGr, group web site: http://kclmmag.org).

Paul Aljabar is a mathematician who enjoys using computer programming to address health and biomedical problems. He taught mathematics in secondary (high) schools in London for twelve years before taking up a research career. Since then, his work has focused on the analysis of large collections of medical images for a range of applications, for example in order to build anatomical atlases or distinguish normal from pathological physiology. As described above, Paul and Andrew developed this book and its teaching materials together whilst teaching Biomedical Engineering undergraduates. Paul has taught on a range of undergraduate and graduate programs where the analysis and interpretation of medical and biomedical data may be carried out through modeling, programming, and the application of methods that are also used in his research (Google Scholar: https://goo.gl/jAgPru).

Preface

AIMS AND MOTIVATION

This book aims to teach the fundamental concepts of computer programming to students and researchers who are studying or working in the biomedical sciences. The book and associated materials grew out of the authors' experience of teaching a first year undergraduate module on computer programming to biomedical engineering students at King's College London for a number of years. Such students have a strong interest in the biomedical applications of computing, and we felt that it was important to provide not only a concise introduction to the key concepts of programming, but also to make these concepts relevant through the use of biomedical examples and case studies.

The book is primarily aimed at undergraduate students of biomedical sciences, but we hope that it will also act as a concise introduction for students in other disciplines wishing to familiarize themselves with the important field of computer programming.

Our emphasis throughout is on practical skill acquisition. We introduce all concepts accompanied by real code examples and step-by-step explanations. These examples are reinforced by learning activities in the chapter text as well as further exercises for self-assessment at the end of each chapter. Code for all examples, activities and exercises is available for download from the book's web site.

Practical skills will be learned using the MATLAB® software package, focusing on the procedural programming paradigm. All code is compatible with MATLAB® 2016a. Note that we do not intend this text to be an exhaustive coverage of MATLAB® as a software package. Rather, we view it as a useful tool with which to teach computer programming. We are always interested in hearing feedback and suggestions from students (particularly those with an interest in biomedical applications) about material that could be added or omitted.

LEARNING OBJECTIVES

On completion of the book the reader should be able to:

- Analyze problems and apply structured design methods to produce elegant, efficient and well-structured program designs.
- Implement a structured program design in MATLAB®, making use of incremental development approaches.
- Write code that makes good use of MATLAB® programming features, including control structures, functions and advanced data types.
- Write MATLAB® code to read in medical data from files and write data to files.
- Write MATLAB® code that is efficient and robust to errors in input data.
- Write MATLAB® code to analyze and visualize medical data, including images.

HOW TO USE THIS BOOK

The book is primarily intended as a companion text for a taught course on computer programming. Each chapter starts with clear learning objectives and the students are helped to meet these objectives through the text, examples and activities that follow. The objectives associated with each example or activity are clearly indicated in the text. Exercises at the end of each chapter enable students to self-assess whether they have met the learning objectives.

The book comes with an on-line teaching pack containing teaching materials for a taught undergraduate module to accompany the book (sample lesson plans, lecture slides and suggestions for coursework projects), as well as MATLAB® code and data files for all examples, activities and exercises in the book. All materials are available from the book's web site at: http://textbooks.elsevier.com/web/Manuals.aspx?isbn=9780128122037.

The book consists of 12 chapters:

- *Chapter 1 – Introduction to Computer Programming and MATLAB®*: This chapter introduces the reader to the fundamental concepts of computer programming and provides a hands-on introduction to the MATLAB® software package. At this stage we go into just enough detail to enable the reader to start using MATLAB® to perform simple operations and produce simple visualizations. We also start to show the reader how to create and debug simple computer programs using MATLAB®.
- *Chapter 2 – Control Structures*: Here, we introduce the concept of control structures, which can be seen as the basic building blocks of computer programs. We show how they can be used to create more complex code resulting in more powerful and flexible programs. Conditional and iteration control structures are introduced and we describe the MATLAB® commands that are provided to implement them.

- *Chapter 3 – Functions*: In this chapter we introduce functions, an important means for programmers to split large programs into smaller modules. As we start to tackle more complex problems, functions are an essential part of the programmer's toolbox. We show how data can be passed into and returned from functions and also introduce special topics such as variable scope and recursion.
- *Chapter 4 – Program Development and Testing*: Next, we cover the important (but often overlooked) topic of how to go about developing a program and checking that it does what it is supposed to do. We introduce the concept of incremental development, in which we always maintain a working version of the program, even it is trivial, and then gradually add more functionality to it. We also give further detail on debugging and describe how we can make our programs robust to errors.
- *Chapter 5 – Data Types*: Up until this point, we will have been dealing with programs that use fairly simple data, such as numbers or letters. In this chapter we explore data types in more depth, investigating how to find out the type of a variable, how to convert between types, as well as introducing some more complex (but powerful) data types such as cell arrays.
- *Chapter 6 – File Input/Output*: Many biomedical applications involve reading data from external files and/or writing data to them. In this chapter we describe the commands that MATLAB® provides for these operations. We describe the different types of file that might be encountered, and cover the commands intended for use on each type.
- *Chapter 7 – Program Design*: While it is necessary to know how to write working code, when tackling larger and more complex problems, it becomes important to be able to break a program down into a number of simpler modules, modules that can interact with each other by passing data. Such an approach to coding makes the process quicker and the resulting code is likely to contain fewer errors. Program design refers to the process of deciding which modules to define and how they interact. We illustrate this process using the commonly used approach of top-down step-wise refinement and we show how structure charts and pseudocode can be used to document a design.
- *Chapter 8 – Visualization*: In this chapter we go into more detail about how to produce visualizations of data. We extend the basic coverage introduced in Chapter 1 and describe how to create plots containing multiple datasets, and plots that visualize multivariate data as well as images.
- *Chapter 9 – Code Efficiency*: There are typically several different ways that a computer program can solve a given problem, and some may be more (or less) efficient. Here we consider how the way in which a program solves a problem affects it efficiency, in terms of time and the computer's memory. We show how efficiency can be measured and discuss ways in which program efficiency can be improved.

- *Chapter 10 – Signal and Image Processing*: Many biomedical applications can involve processing of large amounts of data. This chapter describes techniques for processing signals (i.e. one-dimensional data) or images (two-dimensional or even three- or four-dimensional data). We introduce the powerful techniques of filtering and convolution and demonstrate how they can be applied to one- or higher-dimensional data. We return to the topic of images (first introduced in Chapter 8) and further show how pipelines may be constructed from simple operations to build more complex and powerful image-processing algorithms.
- *Chapter 11 – Graphical User Interfaces*: As well as the functionality of a program (i.e. what it does) it is important to consider how it will interact with its users. This can be done using a simple command window interface, but more powerful and flexible interfaces can be developed using a GUI (Graphical User Interface). This chapter describes how GUIs can be created and designed in MATLAB® and how we can write code to control the way in which they interact with the user.
- *Chapter 12 – Statistics*: The final chapter introduces the basics of statistical data analysis with MATLAB®. The chapter does not cover statistical theory in depth, but the intention is rather to describe how the most common statistical operations can be carried out using MATLAB®. Some knowledge of statistics is assumed, although fundamental concepts are briefly reviewed. We cover the fields of descriptive statistics (summarizing and visualizing data) and inferential statistics (making decisions based upon data).

We believe that the first ten chapters of the book are suitable content for an undergraduate module in computer programming for biomedical scientists. The last two chapters are included as more advanced topics which may be useful for students' future work in using MATLAB® as a tool for biomedical problem-solving.

We end each chapter of the book with a brief pen portrait of a famous computer programmer from the past or present. As well as providing a bit of color and interest, we hope that this will help the reader to see themselves as following in the footsteps of a long line of pioneers who have developed the field of computer programming and brought it to where it is today: an intellectually stimulating and incredibly powerful tool to assist people working in a wide range of fields to further the limits of human achievement.

Finally, a word about data. We have made every effort to make our biomedical examples and exercises as realistic as possible. Where possible, we have used real biomedical data. On the occasions where this was not possible, we have tried to generate realistic synthetic data. We apologize if this synthetic data lacks realism or plausibility in any respect, or gives a misleading impression about the application that the exercise deals with. Our intention has been to provide examples that are relevant to those with an interest in biomedical applications, and we accept responsibility for any flaws or errors in the data we provide.

ACKNOWLEDGMENTS

The authors would like to thank the following people for their contributions to the development of this book.

- All biomedical engineering students at King's College London who have taken our Computer Programming module over the past few years, for their constructive feedback which has improved the quality of our teaching materials that formed the basis for this book. Also, for the same reasons, all teaching assistants and lecturers who have contributed to the delivery the module, in particular Alberto Gomez.
- Dr. Adam Shortland of the One Small Step Gait Laboratory at Guy's Hospital, London, for insightful discussions and providing the data for a number of activities and exercises throughout the book: Activity 1.8, Exercise 2.8, Exercise 3.6, Activity 5.8, Activity 8.2, and Exercise 8.4.
- From the publishers, Elsevier: Tim Pitts for prompting us to write the book in the first place, and Anna Valutkevich for encouraging us to produce it on time.

CHAPTER 1

Introduction to Computer Programming and MATLAB

LEARNING OBJECTIVES

At the end of this chapter you should be able to:

O1.A Describe the broad categories of programming languages and the difference between compiled and interpreted languages

O1.B Use and manipulate the MATLAB environment

O1.C Form simple expressions using scalars, arrays, variables, built-in functions and assignment in MATLAB

O1.D Describe the different basic data types in MATLAB and be able to determine the type of a MATLAB variable

O1.E Perform simple input/output operations in MATLAB

O1.F Perform basic data visualization in MATLAB by plotting 2-D graphs and fitting curves to the plotted data

O1.G Form and manipulate matrices in MATLAB

O1.H Write MATLAB scripts, add meaningful and concise comments to them and use the debugger and code analyzer to identify and fix coding errors

1.1 INTRODUCTION

The processing power of digital computers has increased massively since their invention in the 1950s. Today's computers are about a trillion times more powerful than those available 60 years ago. Likewise, the use of computers (and technology in general) in biology and medicine (commonly termed *biomedical engineering*) has expanded hugely over the same time period. Today, diagnosis and treatment of many diseases is reliant on the use of technology such as imaging scanners, robotic surgeons and cardiac pacemakers. Computers and computer programming are at the heart of this technology.

The simultaneous rise in computing power and the biomedical use of technology is not a coincidence: advances in technology often drive advances in biology and medicine, enabling more sensitive diagnostic procedures and more accurate and effective treatments to be developed. In recent years, the

MATLAB Programming for Biomedical Engineers and Scientists. DOI: 10.1016/B978-0-12-812203-7.00001-X ever-closer relationship between technology and medicine has resulted in the emergence of a new breed of medical professional: someone who both understands the biological and medical challenges but also has the skills needed to apply technology to address them and to exploit any opportunities that arise. If this sounds like you (or perhaps the person you would like to be), then this book is for you. Over the next twelve chapters we hope to introduce the fundamental concepts of computer programming but with a strong biomedical focus, which we hope will give you the skills you need to tackle the increasingly complex problems faced in modern medicine.

In this first chapter of the book we will introduce some fundamental concepts about computer programming, and then focus on familiarizing ourselves with the *MATLAB* software application. This is the software that will be used throughout the book to learn how to develop computer programs.

1.1.1 Computer Programming

The development of the modern digital computer marked a step change in the way that people use machines. Until then, mechanical and electrical machines had been proposed and built to perform *specific operations*, such as solving differential equations. In contrast, the modern digital computer is a *general purpose machine* that can perform any calculation that is computable. The way in which we instruct it what to do is by using computer programming.

We can make a number of distinctions between different computer programming languages depending on how we write and use the programs. Firstly, a useful distinction is between an *interpreted* language and a *compiled* language. A program written in an interpreted language can be run (or *executed*) immediately after it has been written. For compiled languages, we need to perform an intermediate step, called *compilation*. Compiling a program simply means translating it into a language that the computer can understand (known as *machine code*). Once this is done, the computer can execute our program. With an interpreted language this 'translation' is done in real-time as the program is executed and a compilation step is not necessary.

In this book we will be learning how to program computers using a software application called *MATLAB*. MATLAB is (usually) an *interpreted* language. Although it is also possible to use it as a compiled language, we will be using it solely as an interpreted language.

Another distinction we can make between different programming languages relates to how the programs are written and what they consist of. Historically, most programming languages have been *procedural* languages. This means that programs consist of a sequence of instructions provided by the programmer. Until the 1990s almost all programming languages were procedural. More recently, an alternative called *object-oriented programming* has gained in popularity. Object-oriented languages allow the programmer to break down the problem into *objects*: self-contained entities that can contain both data and operations to act upon data. Finally, some languages are *declarative*. In principle, when we write a declarative program it does not matter what order we write the program statements in – we are just 'declaring' something about the problem or how it can be solved. The compiler or interpreter will do the rest. Currently, declarative languages are mostly used only for computer science research.

MATLAB is (mostly) a procedural language. It does have some object-oriented features but we will not cover them in this book.

1.1.2 MATLAB

MATLAB is a commercial software package for performing a range of mathematical operations. It was first developed in 1984 and included mostly linear algebra operations (hence the name, MATrix LABoratory). It has since been expanded to include a wide range of functionality including visualization, statistics, and algorithm development using computer programming. It has approximately 1 million current users in academia, research and industry, and has found particular application in the fields of engineering, science and economics. As well as its core functionality, there are a number of specific toolboxes that are available to extend MATLAB's capabilities.

Although MATLAB is a very powerful software package, it is a commercial product. Student licenses can be obtained relatively cheaply, but full licenses are more expensive. If your institution provides a license this is not a problem for you, but if you need to acquire a license yourself, and you are not eligible for a student license, you may want to consider one of the free, open-source alternatives to MATLAB. Some of the most common are:

- *Octave*: Good compatibility with MATLAB, but includes a subset of its functionality (http://www.gnu.org/software/octave);
- *Freemat*: Good compatibility with MATLAB, but includes a subset of its functionality (http://freemat.sourceforge.net);
- *Scilab*: Some differences from MATLAB (http://www.scilab.org/en).

1.2 THE MATLAB ENVIRONMENT

After we start MATLAB, the first thing we see is the MATLAB *environment window*. To begin with, it is important to spend some time familiarizing ourselves with the different components of this window and how to use and manipulate them. Fig. 1.1 shows a screenshot of the MATLAB environment window (if yours does not look exactly like this, don't worry – as we will see shortly it is possible to customize its appearance).

The *command window* is where we enter our MATLAB commands and view the responses MATLAB gives (if any). As we enter commands they will be added to our *command history*. We can always access and repeat old commands through

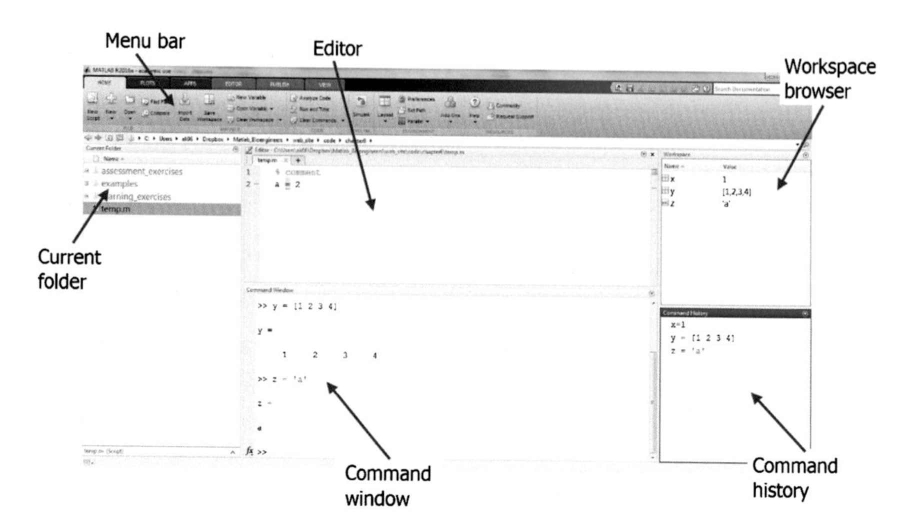

FIGURE 1.1 The MATLAB environment.

the *command history* window. Another way of cycling through old commands in the *command window* is to use the up and down arrow keys. Although the *command window* is the main way in which we instruct MATLAB to perform the operations we want, it is often also possible to execute commands via the *menu bar* at the top of the environment window.

The MATLAB environment window can also be used to navigate through our file system. The *current folder* window shows our current folder and its contents. It can also be used for simple navigation of the file system.

As we use the *command window* to create and manipulate data, these data are added to our *workspace*. The workspace is basically a collection of all of the data that have been created or loaded into MATLAB. The current contents of the workspace are always displayed in the *workspace browser*.

When we start creating scripts to perform more complex MATLAB operations (see Section 1.10), we will use the *editor* window to create and edit our script files. This can be useful if we want to perform the same sequence of operations several times or save a list of operations for future use.

We can customize the MATLAB environment by clicking on the *Layout* button in the menu bar (under the *Home* tab).

■ Activity 1.1

O1.B

Try customizing the MATLAB environment window. For example, switch between *Default* and *Two Column* layouts; hide the *Current Folder* and/or *Workspace* panels, then make them reappear again; try switching the *Command History* between the *Popup* and *Docked* modes and see the effect this has. If you want to get rid of any changes you've made and return to the original layout, then simply select *Layout* → *Default*. ■

1.3 HELP

MATLAB includes extensive documentation and online help. At any time we can access this through the menu system (*Help* → *Documentation* under the *Home* tab). We can also access help in the command window (type `doc` or `help` followed by the command we want information about). All documentation is also available online so we can simply use our favorite search engine and enter the keyword "MATLAB" together with some keywords related to the task we wish to perform.

■ Activity 1.2

O1.B

Try working through some of the simple MATLAB tutorials that are freely available as part of its documentation. For example, select *Help* → *Documentation* then click on *MATLAB* followed by *Getting Started with MATLAB*. Click on *Desktop Basics* under the list of tutorials to learn more about customizing the MATLAB environment. ■

1.4 VARIABLES, ARRAYS AND SIMPLE OPERATIONS

Now we will start to type some simple commands into the command window. The first thing we will do is to create some *variables*. Variables can be thought of as named containers for holding information, often numbers. They can hold information about anything, for example clinical data such as heart rate, blood pressure, height and weight. They are called variables because the values (but not the names) can vary. For example, the following commands create two new variables named `a` and `y` that have values 1 and 3 respectively.

```
>> a = 1
>> y = 3
```

Note that `>>` is the MATLAB *command prompt*, i.e. it will be displayed in the command window to 'prompt' you to enter some commands. We don't need to type this in ourselves, just the text after it.

When we enter these commands into the MATLAB command window we see the new variables appear in the *workspace browser*. These variables can then be used in forming subsequent expressions for evaluation by MATLAB, e.g.

```
>> c = a + y
>> d = a ^ y
>> e = y / 2 - a * 4
```

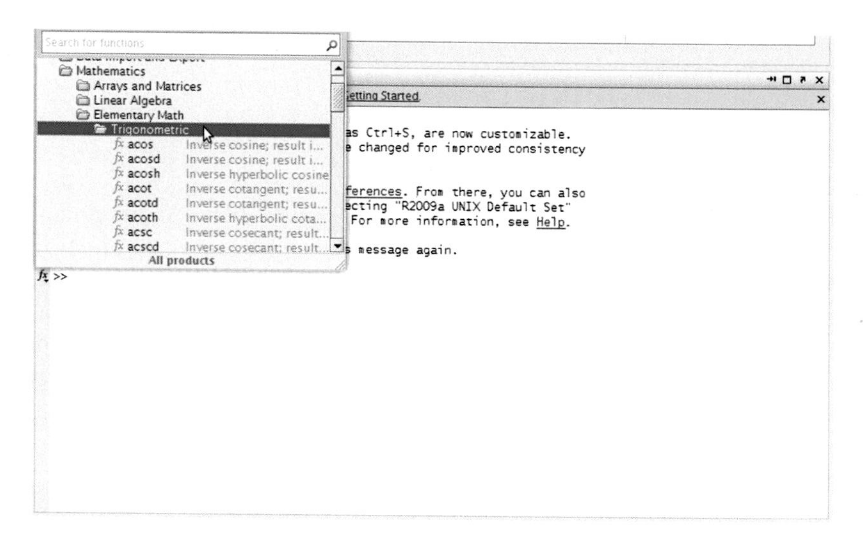

FIGURE 1.2 Accessing MATLAB's built-in functions using the graphical menu.

Again, as we type these commands we create new variables and these are added to the workspace. This time they are called `c`, `d` and `e` (the ^ symbol raises a number to the power of another number, i.e. `a^y` means a^y).

MATLAB contains a large number of built-in functions for operating on numbers or variables. If we know which function we want to use we can just type it into the command window, e.g.

```
>> sin(pi/4)
>> z = tan(a)
```

and the function will be evaluated using *argument(s)* provided in brackets. Note that the name `pi` is a built-in MATLAB variable that contains the value 3.14159...

Alternatively, if we want to browse through all available functions, we can click on the *fx* symbol immediately to the left of the command prompt (see Fig. 1.2). A menu will pop up allowing us to navigate through a range of categories of functions. For example, to find trigonometric functions choose *Mathematics, Elementary Math* and then *Trigonometry*. A list of commonly used mathematical functions that operate on numerical values is given in Table 1.1.

Creating and manipulating arrays: As well as containing single values, variables can also contain lists of values. In this case they are known as *arrays*. It is in the creation and manipulation of arrays that MATLAB can be particularly

Table 1.1 A list of common MATLAB numerical functions

`sin(x)`	Sine (input in radians)	`log10(x)`	Base 10 logarithm
`cos(x)`	Cosine (input in radians)	`sqrt(x)`	Square root
`tan(x)`	Tangent (input in radians)	`abs(x)`	Absolute value
`asin(x)`	Arc sine (result in radians)	`floor(x)`	Round to next smallest integer
`acos(x)`	Arc cosine (result in radians)	`ceil(x)`	Round to next largest integer
`atan(x)`	Arc tangent (result in radians)	`round(x)`	Round to nearest integer
`exp(x)`	exponential of x, i.e. e^x	`fix(x)`	Round to integer toward zero
`log(x)`	Natural logarithm	`mod(x,y)`	Modulus (remainder)

powerful. To illustrate this, consider the following commands. First, we create a sample array variable,

```
>> d = [1 3 4 2 5];
```

We define arrays by listing all values of the array, separated by spaces, enclosed by square brackets. Alternatively, we can separate the values by commas, so the following two commands are equivalent,

```
>> d = [1 3 4 2 5];
>> d = [1,2,3,4,5];
```

Note that adding a semi-colon at the end of the command causes MATLAB to suppress its response to the command. The variable will be created and added to the workspace, but MATLAB won't display anything in the command window. Note also that by entering the above command we will overwrite our previous definition of `d = a^y`.

Now, suppose that we wish to compute the value of each array element raised to the power of 3. We simply type the following command,

```
>> d.^3
```

If we perform an arithmetic operation like this on an array variable MATLAB will automatically apply the operation to each element of the array. This can make processing of large amounts of data very easy. However, note the '`.`' symbol before the caret (`^`). This tells MATLAB to apply the `^` operator to each element *individually*, or *element-wise*. Without the '`.`', common arithmetic operators will be interpreted differently and can have special meanings when applied to arrays. We will see this later in Section 1.9.

Note how spaces are treated in MATLAB. In particular, spaces are used to separate elements of arrays, as in the command `d = [1 3 4 2 5]` we saw above.

Otherwise, when evaluating expressions spaces are ignored by MATLAB, e.g. the following two commands are identical:

```
>> d.^3
>> d .^ 3
```

MATLAB is, however, *case-sensitive*: the letters `d` and `D` represent different variables in the MATLAB workspace.

It is common to create a large array of numbers with values that are regularly spaced. In this case there is a shorthand way to enter the values, as an alternative to typing in each value individually. The '`:`' operator allows us to create an array by specifying its first value, the spacing between adjacent values, and the last value. For example, try typing the following,

```
>> x = 0:0.25:3
```

and look at the resulting array. It consists of values separated by 0.25, starting with 0 and ending with 3. The colon operator is a much easier way to create arrays with a large number of elements.

An alternative way to create such regularly spaced arrays is to use the MATLAB `linspace` command. Consider the following code

```
>> x = linspace(0, 3, 13)
```

which has the same effect as the colon operator example shown above. The three arguments passed to the `linspace` function (inside the brackets) are the first value, the last value and the number of values in the array.

Accessing array elements: So far we have seen how to create arrays and to perform operations on the array elements. Often we will then want to access individual array elements. In MATLAB, this can be done using round brackets, e.g.

```
>> a = x(3);
```

This command accesses the third element of the array `x` and assigns it to the variable a. We also refer to this operation as *array indexing*, and the value in brackets (`3` in this case) is known as the array *index*. Similarly we can use array indexing to assign values to individual array elements without affecting the rest of the array,

```
>> x(3) = 10
```

Array functions: We have already seen that MATLAB has a number of built-in functions for operating on numbers and numerical variables. There are also a

Table 1.2 A list of common MATLAB array functions

`max(x)`	Maximum value	`cross(x,y)`	Vector cross product
`min(x)`	Minimum value	`dot(x,y)`	Vector dot product
`range(x)`	Range of values	`disp(x)`	Display an array
`sum(x)`	Sum of all values	`mean(x)`	Mean value of array
`prod(x)`	Product of all values	`std(x)`	Standard deviation of array
`length(x)`	Length of array		

number of built-in functions specifically for operating on arrays. A list of common MATLAB array operations is given in Table 1.2. For example, try typing the following,

```
>> a = [1 5 3 9 11 3 23 6 13 14 3 9];
>> max(a)
>> min(a)
>> sum(a)
```

In addition to the array functions listed in Table 1.2, all of the numerical functions summarized in Table 1.1 can be used with array arguments. In this case, the function will be applied to each array element separately and the answers combined into an array result. For example, try typing the following commands,

```
>> x = [10 3 2 7 9 12];
>> sqrt(x)
>> y = [-23 12 0 1 -1 43 -100];
>> abs(y)
```

■ Activity 1.3

O1.C

The probability of getting cancer has been shown to have an approximately linear relationship with the number of cigarettes smoked. Enter the array `[5 2 14 7 9 11]` into MATLAB, which represents the number of cigarettes smoked each by a group of six smokers. Assuming that the relationship between smoking and cancer risk is

cancer probability $= 0.23 \times$ number of cigarettes $+ 0.8$,

perform the following operations:

1. Compute the probability of each of the six smokers getting cancer.
2. Replace the second element of the array with the number 12, then recompute the probabilities for the entire array.
3. Add a new number 20 to the end of the array, and recompute all probabilities again.
   (Hint: In MATLAB the word `end`*, when used as an array index, refers to the last element of the array.)* ■

1.5 DATA TYPES

So far, all variables and values we have used have been numeric. Although MATLAB is mostly used for numerical calculations, it is possible, and sometimes very useful, to manipulate data of other types. MATLAB allows the following basic data types (the names in brackets are the terms MATLAB uses for the type):

- Floating point, e.g. 1.234 (`single`, `double` depending on the precision required)
- Integer, e.g. 3 (`int8`, `int16`, `int32`, `int64` depending on the number of bits to be used to store the number)
- Character, e.g. 'a' (`char`)
- Boolean, e.g. true (`logical`)

Other, more advanced types are also possible but we won't discuss them just yet. (We'll return to the concept of data types in Chapter 5.) Note that numeric values can be either floating point or integer types. In particular, unless we explicitly specify otherwise, all numeric values in MATLAB will be `double` precision floating point values. You can always find out the type of a variable(s) using the `whos` command. If we just type `whos` at the command line, we obtain a list of all variables in the workspace and their types. If we give specific variable names, it will only display the types of those given.

■ Example 1.1

O1.D

For example, try typing the following in the command window:

```
>> a = 1
>> b = 'x'
>> c = false
>> whos
>> d = [1 2 3 4]
>> e = ['a' 'b' 'c' 'd']
>> f = ['abcd']
>> whos d e f
```

Note how the *character* type is indicated by enclosing characters in single quotes (MATLAB does not recognize double quotes). Note also that the definitions of the variables `e` and `f` are equivalent: the form used in assigning to `f` is short-hand for that used in assigning to `e`. We will see a practical use for arrays of characters (also called *strings*) in the next section. ■

Sometimes we want to convert between different data types. We will discuss this topic in detail in Section 5.8.3 but for now we introduce an example usage of type conversion. When using the `disp` function to display an array (see Table 1.2) we may want to mix different types, e.g. a string and a number. Normally in MATLAB this is not possible because arrays can only contain data of a single type. However, we can get around this by converting the types of some of our data as shown below.

```
>> sys = 100;
>> str = ['Systolic B.P. = ' num2str(sys)];
>> disp(str);
```

Here, we want to mix string data with a numeric value, representing systolic blood pressure. We use the built-in MATLAB function `num2str` to convert the numeric value into its string equivalent, and then combine the string that forms the start of the message with the converted numeric data. This results in a single string (character array) which is assigned to `str`, which is then displayed using the `disp` command.

■ Activity 1.4

Type the following in the MATLAB command window:

O1.D

```
>> heart_rate = 65
>> blood_sugar = 6.2
>> name = 'Joe Bloggs'
>> rate = num2str(heart_rate)
```

Predict what the data types of the four variables will be. Verify your predictions by using the `whos` command. ■

1.6 LOADING AND SAVING DATA

When using MATLAB for biomedical problems, we may be working with large amounts of data that are saved in an external file. Therefore, it is important to know how to load data from, and save data to, external files.

When writing data items to a text file, we normally need a special character called a *delimiter* that is used to separate one data item from the next. We can save data delimited in this way with the `dlmwrite` command as illustrated in the following example,

```
>> f = [1 2 3 4 9 8 7 6 5];
>> dlmwrite('test.txt',f,'\t');
```

Here, the three arguments of `dlmwrite` specify the file name (a string), the variable containing the data to be saved, and the delimiter or symbol that will be used in the file to separate the array values (as a single character). In this command the file delimiter is a `'\t'` character, which refers to a *tab* character. We could have used another character as a delimiter, e.g. `','` or `'-'`. The `dlmwrite` command can be useful if we need to write data in a particular format to be read in by another software application.

The load and save commands are probably the most commonly used commands for reading and writing data. The following commands illustrate the use of load and save,

```
>> d = load('test.txt');
>> e = d / 2;
>> save('newdata.mat', 'd', 'e');
>> save('alldata.mat');
>> clear
>> load('newdata.mat')
```

When using the save command the first argument specifies the name of the file to be saved and any subsequent arguments specify which variable(s) to save inside the file. Omitting the variable name(s) will cause the entire workspace to be saved. save will create a MATLAB-specific binary file containing the variables and their values that can be loaded in again at any time in the future. These binary files are called *MAT* files and it is common to give them the file extension *'.mat'*. The load command can be used to read either text files created by dlmwrite or binary *MAT* files created by save. Note that when using the load command with a text file we should assign the result of the load into a variable for future use (e.g. see the difference between the first and the last commands above). The clear command removes all current variables from the workspace.

An alternative, and easier, way of loading *MAT* files into the workspace is to simply use the MATLAB environment to drag-and-drop the appropriate file icon from the current folder window to the workspace browser (or just double-click on it in the current folder window).

Finally, the *import data* wizard can often be the quickest and easiest way to read data from text files (i.e. not *MAT* files). Selecting the *Import Data* button on the *Home* tab will bring up a wizard in which we can choose the file and interactively specify delimiters and the number of header lines in the file.

We will look again in more detail at the subject of MATLAB file input/output functions in Chapter 6.

■ Activity 1.5

O1.C, O1.E

Continuing with the application from Activity 1.3, create a regularly spaced array of numbers of cigarettes smoked between zero and twenty in steps of two. Use the linear equation provided earlier to compute the probabilities of getting cancer for each of these numbers. Save the probabilities only to a text file called *cancer.txt*. Use the newline character ('\n') as the delimiter. ■

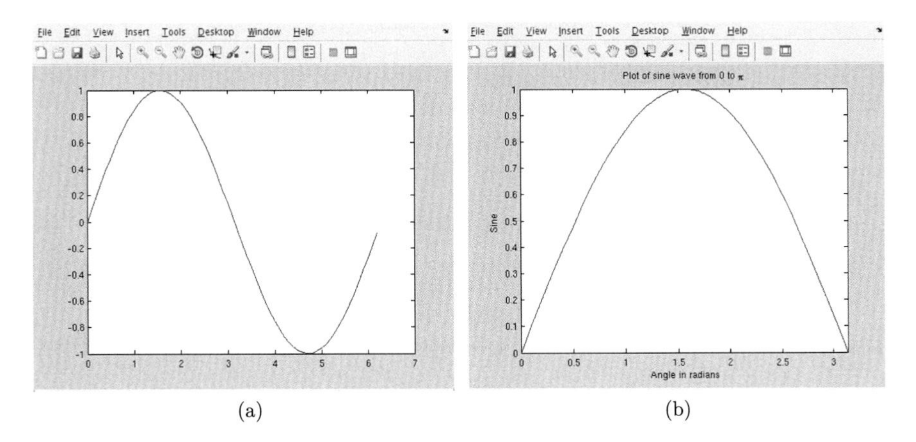

FIGURE 1.3 Plots of a sine wave between (a) 0 and 2π; (b) 0 and π. Plot (b) also has annotations on the axes and a title.

1.7 VISUALIZING DATA

As well as accessing and manipulating data, MATLAB can also be used to visualize it. The `plot` command is used for simple data visualization. Consider the following code,

```
>> x = 0:0.1:2*pi;
>> y = sin(x);
>> plot(x,y,'-b');
```

This should produce the output shown in Fig. 1.3a. The first line in the code uses the colon operator to create an array called `x` containing numbers that start with 0, and go up to 2π in steps of 0.1. This array acts as the x coordinates for the plot. The second line applies the built-in MATLAB `sin` function to every element of `x`. The resulting array acts as the y coordinates for the plot. Finally, the third line of code displays the graph using the `plot` command. The first two arguments are the x and y coordinate arrays (which must be of the same length). The third argument specifies the appearance of the line. Table 1.3 summarizes the different symbols that can be used to control the appearance of the line and the markers. All of these symbols can be included in the third string argument to the `plot` command. Try experimenting with some of these to see the effect they have.

Next, we will add some text to our plot. Try typing the following commands to add a title and labels for the x and y axes respectively.

```
>> title('Plot of sine wave from 0 to 2\pi');
>> xlabel('Angle in radians');
>> ylabel('Sine');
```

Table 1.3 Codes that can be used to specify the appearance of markers and lines in MATLAB plots. They are used in combination and passed to the `plot` command, e.g. `'o:r'` gives a plot with circular markers and a dashed line colored red

	Marker		Line		Color
.	Dot	–	Solid line	k	Black
x	Cross	– –	Dashed line	b	Blue
o	Circle	:	Dotted line	r	Red
d	Diamond	–.	Dash-dot line	g	Green
s	Square			m	Magenta
+	Plus			c	Cyan
*	Asterisk			y	Yellow

If we only wish to visualize the plot for values of `x` between 0 and π, the `axis` command can be used to specify minimum and maximum values for the x and y axes,

```
>> axis([0 pi 0 1]);
```

This should produce the output shown in Fig. 1.3b, i.e. the same plot but zoomed to the x range from 0 to π, and the y range from 0 to 1.

Sometimes we may want to display two or more curves on the same plot. This is possible with the `plot` command as the following code illustrates,

```
>> y2 = cos(x);
>> plot(x,y,'-b', x,y2,'--r');
>> legend('Sine','Cosine');
```

We can display multiple curves by just concatenating groups of the three arguments in the list given to the `plot` command, i.e. x data, y data, line/marker style, etc. The `legend` command adds a legend to identify the different curves. The `legend` command should have one string argument for each curve plotted.

Finally, note that we can close down all figure windows that are currently open by typing

```
>> close all
```

We will return to the subject of data visualization in Chapter 8, and discuss more ways of visualizing multiple datasets in Section 8.2.1.

1.8 CURVE FITTING

As well as plotting data, MATLAB allows us to perform curve fitting in order to help identify relationships between data. As we will see, this can be done in two ways: either by calling the built-in functions `polyfit` and `polyval` from the command window, or by using the menu options on a figure window produced.

■ Example 1.2

To illustrate the use of `polyfit` and `polyval`, consider the following example. When diagnosing heart disease, cardiologists observe and measure the motion of the heart as its beats. Radial myocardial displacements are measured for the left ventricle of a patient's heart using a dynamic magnetic resonance (MR) scan. The data for one segment of the ventricle are contained in a file called *radial.mat*, which can be downloaded from the book's web site. This *MAT* file contains two variables: `radial` represents the radial displacement measurements and `t` represents the time (in milliseconds) of each measurement. The following code will load the data, plot `radial` against `t`, and then fit a cubic polynomial curve to the data.

O1.E, O1.F

```
>> load('radial.mat');

>> plot(t, radial, '-r');
>> title('Radial myocardial displacement versus time');
>> xlabel('Time (ms)')
>> ylabel('Radial displacement (mm)');

>> p = polyfit(t, radial, 3);

>> radial_100ms = polyval(p, 100);
```

Note that the `polyfit` function takes three arguments: the x and y data that we are fitting the curve to, and the order of the polynomial (3 in this case for a cubic polynomial). The value returned (`p`) is an array containing the cubic polynomial coefficients. This is then used as one of the inputs to the `polyval` function. `polyval` computes the value of a polynomial curve at a given value. In the example code, we compute the value of the fitted cubic polynomial (i.e. the predicted radial displacement) at `t` = 100 milliseconds. ■

Alternatively, the same curve fitting can be carried out using the menu options on a figure plot. Try displaying a plot of `t` against `radial` using the code example given above (see Fig. 1.4a), and then selecting the *Tools* menu on the figure window, followed by the *Basic Fitting* option. This opens a new window where we can choose which curve(s) to fit, as well as to predict values using the fitted curve(s) (see Fig. 1.4b). Try experimenting with this functionality.

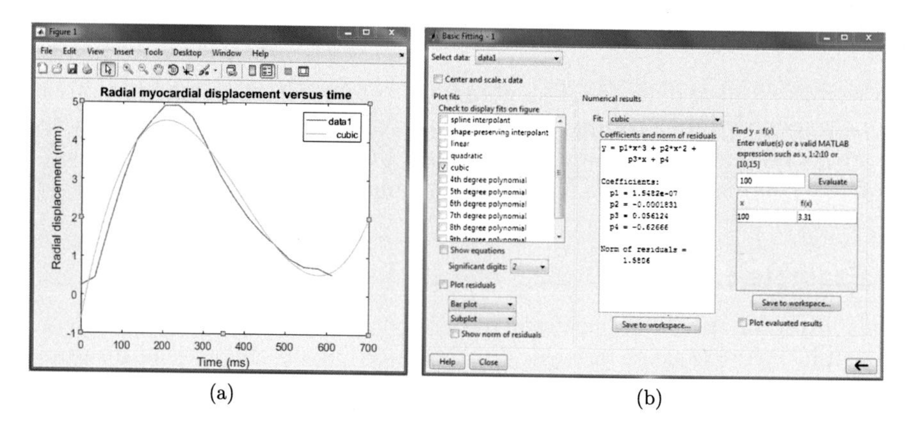

(a) (b)

FIGURE 1.4 Curve fitting using the figure window (a) plot of time against radial myocardial displacement; (b) fitting a cubic polynomial to the data using the graphical user interface.

■ Activity 1.6

O1.C, O1.F

The Injury Severity Score (ISS) is a medical score to assess trauma severity. Data on ISS and hospital stay (in days) have been collected from a number of patients who were admitted to hospital after accidents. The ISS data are `[64 35 50 46 59 41 27 39 66]`, and the length of stay data are `[8 2 5 5 4 3 1 4 6]`. Use MATLAB to plot the relationship between ISS and hospital stay. Then, fit a linear regression line to these data and estimate what length of hospital stay would be expected for a patient with an ISS of 55. ■

1.9 MATRICES

At its simplest, a MATLAB array is a *one-dimensional* (1-D) list of data elements. *Matrices* can also be defined, which are *two-dimensional* (2-D) arrays. In this case we use semi-colons to separate the rows in the matrix, for example:

```
>> a = [1 2; 3 4];
>> b = [2, 4;
1, 3];
```

As with 1-D arrays, the row elements of matrices can be delimited by either spaces or commas. Note also that we can press <RETURN> to move to a new line in the middle of defining an array (either 1-D or 2-D). MATLAB will not process the array definition until we have closed the array with the `]` character.

MATLAB matrix elements can be accessed in the same way as 1-D array elements, except that two indices need to be specified, one index for the row and one index for the column:

Table 1.4 Some MATLAB built-in matrix functions

`eye(n)`	Create `n` by `n` identity matrix
`zeros(m,n)`	Create `m` by `n` matrix of zeros
`ones(m,n)`	Create `m` by `n` matrix of ones
`rand(m,n)`	Create `m` by `n` matrix of random numbers in range [0–1]
`size(a)`	Return the number of rows/columns in the matrix a
`inv(a)`	Return the inverse of a square matrix a
`transpose(a)`	Return the transpose of the matrix a

```
>> a(1,2)
>> b(2,2)
```

If we want to access an entire row or column of a matrix we can use the colon operator[1]

```
>> a(:,2)
>> b(1,:)
```

MATLAB also provides a number of built-in functions specifically intended for use with matrices, and these are summarized in Table 1.4.

■ Activity 1.7

Write MATLAB code to define the following matrices: *O1.G*

$$X = \begin{pmatrix} 4 & 2 & 4 \\ 7 & 6 & 3 \end{pmatrix} \quad Y = \begin{pmatrix} 1 & 2 & 3 \\ 1 & 0 & 0 \end{pmatrix} \quad Z = \begin{pmatrix} 2 & 0 & 9 \\ 1 & 5 & 7 \end{pmatrix}$$

Now perform the following operations:

1. Replace the element in the second row and second column of Z with 1
2. Replace the first row of `X` with `[2 2 2]`
3. Replace the first row of `Y` with the second row of `Z` ■

■ Activity 1.8

Gait analysis is the study of human motion with the aim of assessing and treating individuals with conditions that affect their ability to walk normally. One way of measuring human motion is to attach markers to a person's joints (hip, knee, ankle) and optically track them using cameras whilst the person walks. From this tracking data, important information such as knee *flexion* (i.e. angle) can be derived. The peak knee flexion and *O1.C, O1.E, O1.G*

---

[1] This is a different use of the colon operator from previous uses we have seen (i.e. for generating arrays of numbers).

the range of knee flexion values are indicators that are of particular interest in gait analysis.

The file *knee_flexion.mat* contains flexion data acquired from the left knee of a patient with a neurological disease whilst walking. The file contains one variable called `left_knee_flexion`. This is a matrix in which the first column represents time (in seconds) and the second column represents knee flexion values (in degrees).

Write MATLAB code to load in the *knee_flexion.mat* file and compute and display to the command window the following values:

- The range of knee flexion values (i.e. the difference between the maximum and minimum values).
- The peak knee flexion value.
- The time at which peak knee flexion occurs. *(Hint: Look at the MATLAB documentation for the* `max` *function.)* ■

As well as storing and accessing 2-D data, matrices allow us to perform a range of different *linear algebra* operations. The following code illustrates the use of some of the common MATLAB matrix operations.

```
c = a * b
d = a + b
e = inv(a)
f = transpose(b)
```

Here, the * and + operators automatically perform *matrix* multiplication and addition because their arguments are both matrices. To explicitly request that an operation is carried out *element-wise*, we use a '.' before the operator. For example, note the difference between the following two commands,

```
c = a * b
d = a .* b
```

Here, `d` is the result of element-wise multiplication of `a` and `b` whilst `c` is the result of carrying out matrix multiplication.

Note that element-wise addition/subtraction are the same as matrix addition/subtraction so there is no need to use a dot operator with + and −.

■ Activity 1.9

O1.E, O1.G

Write MATLAB commands to define the matrices A, B and C as follows:

$$A = \begin{pmatrix} 1 & 2 & 1 \\ 3 & 2 & 1 \\ 3 & 2 & 4 \end{pmatrix} \quad B = \begin{pmatrix} 4 & 4 & 4 \\ 2 & 2 & 2 \\ 1 & 1 & 1 \end{pmatrix} \quad C = \begin{pmatrix} 1 & 0 & 1 \\ 2 & 1 & 3 \\ 1 & 0 & 1 \end{pmatrix}$$

Now compute the following:

1. $D = (AB)C$
2. $E = A(BC)$
3. $F = (A + B)C$
4. $G = A + BC$

Save all variables to a MATLAB *MAT* file. ■

Finally, matrices can also be used to solve systems of linear equations such as:

$$5x_1 + x_2 = 5$$
$$6x_1 + 3x_2 = 9$$

We approach this problem by first expressing the equations in matrix form, i.e.

$$\begin{pmatrix} 5 & 1 \\ 6 & 3 \end{pmatrix} \begin{pmatrix} x_1 \\ x_2 \end{pmatrix} = \begin{pmatrix} 5 \\ 9 \end{pmatrix}$$

Then, the system of equations can be solved by premultiplying by the inverse of the first matrix to solve for x_1 and x_2:

$$\begin{pmatrix} x_1 \\ x_2 \end{pmatrix} = \begin{pmatrix} 5 & 1 \\ 6 & 3 \end{pmatrix}^{-1} \begin{pmatrix} 5 \\ 9 \end{pmatrix}$$

■ Example 1.3

The following MATLAB code implements the solution to the system of linear equations given above. Enter the commands in MATLAB to find the values of x_1 and x_2. *O1.G*

```
M = [5 1; 6 3];
y = [5; 9];
x = inv(M) * y
```

■

The following activity is one that may be solved using simultaneous equations and matrices.

■ Activity 1.10

A doctor wishes to administer 2 ml of a drug in a solution which has a 10% concentration. However, only concentrations of 5% and 20% of the drug are available. Use MATLAB to determine how much of the 5% and 20% solutions the doctor should mix to produce 2 ml of 10% concentration solution. ■ *O1.G*

1.10 MATLAB SCRIPTS

As we start to write slightly longer pieces of MATLAB code, we will find it useful to be able to save sequences of commands for later reuse. MATLAB *scripts* (or script *m*-files) are the way to do this. To create a new MATLAB script file we can use the *editor* window. First, we click on the *New* button (under the *Home* tab) and choose *Script*. This will create a blank file. We can now type our MATLAB commands into this file. To save the file, we click on the *Save* button, (under the *Editor* tab). The first time that we save the file we will be prompted for a file name. It is important to save MATLAB script files with the file extension '*.m*' (hence the name *m*-file). To run the commands in our script, we click on the *run* icon, .

■ Example 1.4

O1.C, O1.F, O1.H

The code listing shown below is an extended version of Example 1.2 (plotting and fitting a curve to radial displacements of the left ventricle of the heart). We can enter this code into a MATLAB script *m*-file, save it with the file name *example_1_4.m* and then run the script as described above. (Note that we do *not* include the MATLAB command prompt in this listing because we are entering the code into the editor window not the command window.)

```
load('radial.mat');

p = polyfit(t, radial, 3);

t_curve=linspace(min(t), max(t), 20);
radial_curve=polyval(p, t_curve);

plot(t, radial, 'or', t_curve, radial_curve, '-b');
title('Radial myocardial displacement versus time');
xlabel('Time (ms)')
ylabel('Radial displacement (mm)');
legend('Displacement data', 'Fitted cubic polynomial')
```

Compare this code to that given in Example 1.2. What does the extra code in this script do? ■

■ Example 1.5

O1.C, O1.F, O1.H

We now add the following code to Example 1.4.

```
[v, end_diastole_frame] = max(radial);
[v, end_systole_frame] = min(radial);

est_end_dia = polyval(p, t(end_diastole_frame));
est_end_sys = polyval(p, t(end_systole_frame));

radial_difference = est_end_dia - est_end_sys;
```

The code has been extended to compute the difference between the maximum and minimum displacements using the fitted curve. Note how we use the optional second return value of the `max` and `min` functions to get the array *indices* of the maximum and minimum displacements. These represent the *end diastole* and *end systole* frames respectively (i.e. at maximum relaxation and maximum contraction respectively). ■

Sometimes we may write a script in which a line of code becomes quite long. For MATLAB, there is no problem with having long lines of code. The MATLAB interpreter will execute them just like any other line. But when it comes to reading and understanding a script, very long lines can make things difficult. In such cases, we can use the special MATLAB symbol "..." (i.e. three dots). This tells MATLAB that we wish to continue the current command on the next line. For instance, we can add the following code to the end of the listing from Example 1.5.

```
disp(['Radial difference between systole/diastole = ' ...
num2str(radial_difference) 'mm']);
```

■ Activity 1.11

O1.C, O1.E, O1.H

Write a MATLAB script *m*-file to read in a height (in meters) and a weight (in kilograms) from the keyboard, then compute and display the body mass index (BMI), according to the formula

$\mathrm{BMI} = \mathrm{mass}/\mathrm{height}^2$.

(Hint: Look at the MATLAB documentation for the `input` *command.)* ■

■ Activity 1.12

O1.C, O1.E, O1.H

Important information about the function of the heart can be learned by measuring the volume of its chambers over the cardiac cycle. The left ventricle is of particular interest as it acts to pump blood around the body. As a very simple approximation, the left ventricle can be considered to be a half-ellipsoid. The equation for the volume of an ellipsoid is

$$\frac{4}{3}\pi abc$$

where a, b and c are the radii of the ellipsoid along its three axes. Estimates of these radii can be made using an imaging device such as an *echocardiography* (i.e. cardiac ultrasound) machine.

Write a MATLAB script *m*-file to read in values from the keyboard for the three radii of a left ventricle (in mm) and estimate its volume by approximating it as a half-ellipsoid. The volume should be reported in mm^3 and also in mL ($1000\ \mathrm{mm}^3 = 1$ mL). You can test your script using typical values for the left ventricle radii at diastole (i.e. maximum relaxation) which are approximately 20–30 mm for the two short axes and 60–90 mm for the long axis. The range of volumes that correspond to these typical dimensions is 50–170 mL. ■

1.11 COMMENTS

When writing MATLAB scripts it is possible to add *comments* to our code. A comment is a piece of text that will be ignored by MATLAB when running our script. Comments in MATLAB are specified by the '%' symbol: any text after a '%' symbol on any line of a script will not be interpreted by MATLAB. It is good practice to add comments to our code to explain what the code is doing. This is useful if we or other people need to look at our code in the future with a view to modifying it or fixing errors.

■ Example 1.6

O1.H

Let's add some comments to the script we looked at in Example 1.4 to see how it improves code readability.

```
% load data
load('radial.mat');

% fit curve
p = polyfit(t, radial, 3);

% determine data used to visualise curve
t_curve=linspace(min(t), max(t), 20);
radial_curve=polyval(p, t_curve);

% plot original data and curve
plot(t, radial, 'or', t_curve, radial_curve, '-b');
title('Radial myocardial displacement versus time');
xlabel('Time (ms)')
ylabel('Radial displacement (mm)');
legend('Displacement data', 'Fitted cubic polynomial')
```

■

■ Activity 1.13

O1.H

Go through some of the example and activity solution files and add meaningful but concise comments to help the readability of the code. ■

1.12 DEBUGGING

Unfortunately, often the programs we write will not work the first time we run them. If they have errors, or *bugs*, in them, we must first *debug* them. MATLAB has two useful features to help us in the debugging process.

1.12.1 MATLAB Debugger

MATLAB has a built-in debugger to help us identify and fix bugs in our code. One of the most useful features of the MATLAB debugger is the ability to set

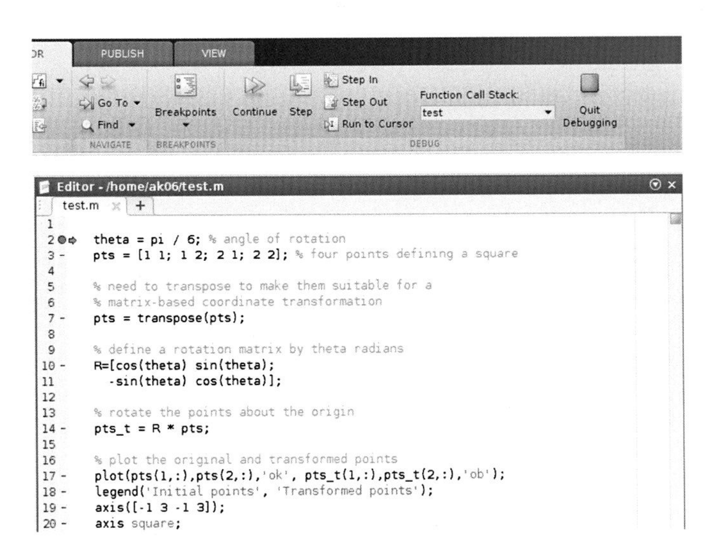

FIGURE 1.5 The MATLAB debugger.

breakpoints. A breakpoint at a particular line in a program will cause execution of the program to automatically stop at that line. We can then inspect the values of variables to check that they are as we expect, and step through the code line by line. To set a breakpoint on a particular line, in the *editor* window left-click once in the gray vertical bar just to the right of the line number. To confirm that the breakpoint has been created MATLAB will display a small red circle in the gray bar (see Fig. 1.5, line 2). Now, when we click on the *run* icon to start the script, execution will stop at the line with the breakpoint. This fact will be indicated by a green arrow just to the right of the breakpoint: (see Fig. 1.5).

Once execution has been stopped, we are free to inspect variable values as we normally would in MATLAB, i.e. by typing in the command window or by looking at the workspace browser. To continue executing the program, there are a number of options, each available from an icon in the *Debug* section of the menu under the *Editor* tab:

- *Continue*: Execute the program from the breakpoint until program termination.
- *Step*: Execute the current line of code, then pause again.
- *Step in*: If the current line is a function, step into that function.
- *Step out*: If we are currently inside a function, step out of the function.
- *Run to cursor*: Execute the program until the line where the cursor is positioned is reached, then pause.
- *Quit debugging*: Exit the debugger.

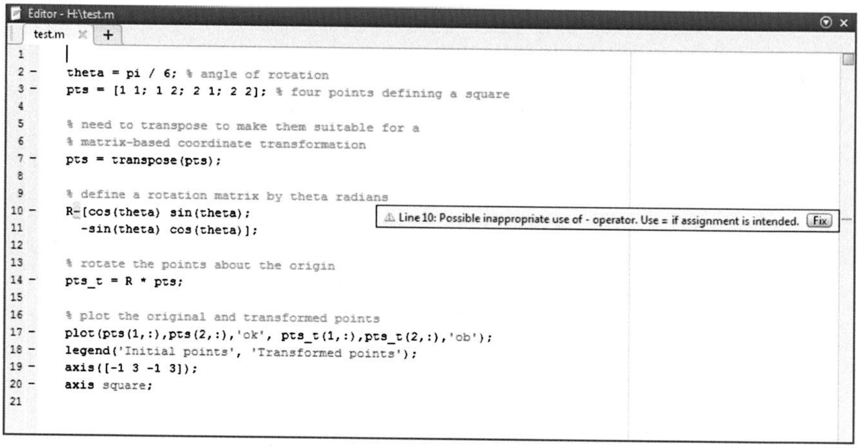

```
Editor - H:\test.m
test.m  ×  +
1
2 -   theta = pi / 6; % angle of rotation
3 -   pts = [1 1; 1 2; 2 1; 2 2]; % four points defining a square
4
5     % need to transpose to make them suitable for a
6     % matrix-based coordinate transformation
7 -   pts = transpose(pts);
8
9     % define a rotation matrix by theta radians
10 -  R-[cos(theta) sin(theta);
11      -sin(theta) cos(theta)];
12
13    % rotate the points about the origin
14 -  pts_t = R * pts;
15
16    % plot the original and transformed points
17 -  plot(pts(1,:),pts(2,:),'ok', pts_t(1,:),pts_t(2,:),'ob');
18 -  legend('Initial points', 'Transformed points');
19 -  axis([-1 3 -1 3]);
20 -  axis square;
21
```

Line 10: Possible inappropriate use of - operator. Use = if assignment is intended. Fix

FIGURE 1.6 The MATLAB code analyzer.

1.12.2 MATLAB Code Analyzer

The second MATLAB feature that is useful for finding and fixing errors is the *code analyzer*. This can be used to automatically check our code for possible problems. It can be run in two ways: either on demand by requesting a code analyzer report, or in real time in the editor window.

Your MATLAB installation is probably set up to analyze your code in real time (i.e. whenever you save a script file in the editor window). To view or change this preference click on *Preferences* under the *Home* tab, and select *MATLAB → Code Analyzer*. Then, either check or uncheck the option labeled *Enable integrated warning and error messages*.

We can see an example of the real time code analyzer in Fig. 1.6. A potential error has been spotted on line 10 and highlighted. It shows that a minus sign has been used where an assignment was expected. A small orange horizontal bar to the right of the text window indicates a warning (a red bar would indicate an error). When the user hovers their mouse pointer over this bar a message pops up, in this case warning of a "possible inappropriate use of - operator". Clicking on the *Fix* button will perform the fix recommended by the code analyzer, in this case to replace the '–' with a '='.

Alternatively, if we don't want such real time warning and error messages, we can change the preference described above, and just request a code analyzer report whenever we want one. To do this, click on the symbol at the top right of the editor window, and select *Show Code Analyzer Report*.

We can also request a code analyzer report from the command window by using either the command `checkcode` or the command `mlintrpt`. In each case, we need to specify the file that we want analyzed:

```
>> checkcode('test.m')
L 10 (C 2): Possible inappropriate use of - operator.
Use = if assignment is intended.
```

The command `mlintrpt('test.m')` can also be invoked from the command window and will produce the same report but it will be given in a separate window.

We will return to the important topic of code debugging in Chapter 4.

■ Activity 1.14

O1.C, O1.E, O1.G, O1.H

The MATLAB *m*-file shown below is intended to read a matrix from a text file and display how many coefficients the matrix has. However, the code does not currently work correctly. Use the MATLAB code analyzer and debugger to identify the bug(s) and fix the code.

```
% clear workspace
clear

% load matrix from text file
m = load('m.txt');

% compute number of coefficients
s = size(m);
n = s^2;

% display answer
disp(['The matrix has ' num2str(n) ' elements']);
```

(This file *activity_1_14.m*, and the test file *m.txt* are available for download from the book's web site.) ■

1.13 SUMMARY

Learning how to program a computer opens up a huge range of possibilities: in principle, we can write a program to perform any computable operation. Programming languages are the means with which we instruct the computer what to do. There are several ways of distinguishing between different types of programming language, e.g. interpreted/compiled, procedural/object-oriented/declarative. In this book we will learn procedural programming using MATLAB, which is normally used as an interpreted programming language.

As well as being a programming language, MATLAB is a powerful software package that implements a wide range of mathematical operations. It can be used to create and assign values to variables, which can have different types, such as numbers, characters, Booleans or 1-D or 2-D arrays (matrices). Basic arithmetic operations and a large number of built-in functions can be applied to values as well as to arrays and matrices. MATLAB can also be used to visualize data by plotting one array of values against another and also to fit curves to the data.

MATLAB scripts (or script *m*-files) allow us to save a sequence of MATLAB commands for repeated use. It is a good idea to add comments to scripts to make them easier to understand. The MATLAB debugger and code analyzer can be used to track down and fix bugs in our code.

1.14 FURTHER RESOURCES

- The MATLAB documentation contains extensive information and examples about all commands and built-in functions. Just type `doc` followed by the command/function name at the command window. This documentation is also available on-line at the Mathworks website.
- The Mathworks website contains a number of useful video tutorials: www.mathworks.com/products/matlab/videos.html.
- The File Exchange section of the same website can be useful for downloading functions and MATLAB code that have been shared by others: www.mathworks.com/matlabcentral/fileexchange.

EXERCISES

■ Exercise 1.1

O1.B, O1.C, O1.D

First, write a command to clear the MATLAB workspace. Then, create the following variables:

1. A variable called `a` which has the character value 'q'.
2. A variable called `b` which has the Boolean value `true`.
3. A variable called `c` which is an array of integers between 1 and 10.
4. A variable called `d` which is an array of characters with the values 'h', 'e', 'l', 'l', 'o'.

Now find out the data types of all variables you just created. ■

■ Exercise 1.2

O1.C

A group of patients have had their heights in centimeters measured as: 159, 185, 170, 169, 163, 180, 177, 172, 168 and 175. The same patients' weights in kilograms are: 59, 88, 75, 77, 68, 90, 93, 76, 70 and 82. Use MATLAB to compute and display the mean and standard deviation of the patients' heights and weights. ■

■ Exercise 1.3

O1.C

Let `a = [1 3 1 2]` and `b = [7 10 3 11]`.

1. Sum all elements in `a` and add the result to each element of `b`.

2. Raise each element of `b` to the power of the corresponding element of `a`.
3. Divide each element of `b` by 4. ■

■ Exercise 1.4

Write MATLAB commands to compute the *sine, cosine* and *tangent* of an array of numbers between 0 and 2π (in steps of 0.1). Save the original input array and all three output arrays to a single *MAT* file and then clear the workspace. ■

O1.C, O1.E

■ Exercise 1.5

Write a MATLAB script *m*-file to read in three floating point values from the user (a, b and c), which represent the lengths of three sides of a triangle. Then compute the area of the triangle according to the equation:

O1.C, O1.E, O1.H

$$\text{Area} = \sqrt{s(s-a)(s-b)(s-c)}$$

where s is half of the sum of the three sides. ■

■ Exercise 1.6

A quick way of determining if any integer up to 90 is divisible by 9 is to sum the digits of the number: if the sum is equal to 9 then the original number is also divisible by 9. Write a MATLAB script *m*-file to verify the rule, i.e. create an array of all multiples of 9 between 9 and 90, and compute the sums of their two digits.

O1.C, O1.H

(Hint: Look at the MATLAB documentation for the `idivide` *and* `mod` *commands. These can be used to find the result and remainder after integer division, so if you use them with a denominator of 10 they will return the two digits of a decimal number. Because these two commands only work with integer values you will also need to know how to convert from floating point numbers to integers: type* `doc int16`*.)* ■

■ Exercise 1.7

Write a MATLAB script *m*-file to plot curves of the *sine, cosine* and *tangent* functions between 0 and 2π. Your script should display all three curves on the same graph and annotate the plot. Limit the range of the y-axis to between -2 to 2. ■

O1.C, O1.F, O1.H

■ Exercise 1.8

Create the following matrices *using a single line of MATLAB code* each:

O1.G

1. A 4×4 matrix containing all zeros apart from 3s on the diagonal.
2. A 3×4 matrix in which all elements have the value -3. ■

■ Exercise 1.9

O1.G

A magnetic resonance (MR) scanner acquires three image slices with their plane equations in scanner coordinates (in mm) given by:

$$
\begin{aligned}
-5x + 5y + 11z &= 21 \\
-2x - 4y - 2z &= -24 \\
32x + 15y - 29z &= 115
\end{aligned}
$$

It is known that the slices intersect at a point. Find the coordinates of the point. ■

■ Exercise 1.10

O1.E, O1.F, O1.H

Data have been collected for the ages and diastolic/systolic blood pressures of a group of patients who are taking part in a clinical trial. All data are available to you in the files *age.txt*, *dbp.txt* and *sbp.txt*. Write a MATLAB script *m*-file that reads in the data and creates a single figure containing two plots: one of age against systolic blood pressure and one of age against diastolic blood pressure. Your script should annotate the plots. ■

■ Exercise 1.11

O1.C, O1.E, O1.F

Positron emission tomography (PET) imaging involves the patient being injected with a radioactive substance (a *radiotracer*), the concentration of which is then imaged using a scanner. Staff involved in performing PET scans receive some radiation dose as a result of being in the vicinity of the patient whilst the radiotracer is inside their body.

A study is investigating the effective radioactive dose to which staff are exposed as a result of performing PET scanning (e.g. see [1]). Data indicating the effective radiation dose (in microsieverts, μSv) as a function of distance (in meters) from the patient are available in the file *dose_data.mat*. Write a MATLAB script *m*-file that loads in this file and fits a polynomial of an appropriate order to the data. The script should then produce a plot of distance against effective dose and display the fitted curve on the same figure. Finally, the fitted curve should be used to compute the expected effective radiation dose at a distance of 3.25 m. ■

FAMOUS COMPUTER PROGRAMMER: ADA LOVELACE

Augusta Ada King, the Countess of Lovelace (commonly known these days as Ada Lovelace), was an English mathematician and writer. She was born in 1815 and was the daughter of the famous English poet, Lord Byron. Her father left her mother a month after Ada was born, eventually dying of disease in the Greek War of Independence when Ada was eight. Ada's mother encouraged her daughter to become interested in mathematics in an effort to prevent her succumbing to what she saw as Lord Byron's insanity.

As a young adult, Ada Lovelace developed a friendship with the British Mathematician, Charles Babbage. Babbage is well known for designing the Analytical Engine, which is commonly recognized as the world's first mechanical computer. In 1842–1843 Ada translated an article on the Analytical Engine, supplementing it with some notes of her own. It is these notes that are the source of Ada's fame today. They contained what is generally considered to be the first computer algorithm, i.e. a sequence of steps that can be executed by a computing machine. It is this vision of a computer as being more than just a numerical calculator that has had a major influence on the historical development of computers, and has led to her being considered as the world's first computer programmer. However, her program has never been run, because the Analytical Engine was never actually built in its designer's lifetime. (See plan28.org for details of a modern-day project to actually build the Analytical Engine and run Ada's program.)

As well as her early work on computer programming, in 1851 Ada was involved in a disastrous attempt to build a mathematical model for making successful bets. This left her thousands of pounds in debt and resulted in her being blackmailed by one of her colleagues in the attempt, who threatened to tell Ada's husband of her involvement.

Ada Lovelace died of cancer in 1852 at the age of 36. As a result of her importance in the history of computing, one of the early computer programming languages, developed in 1980, was named Ada.

> *"The Analytical Engine weaves algebraic patterns, just as the Jacquard loom weaves flowers and leaves."*
>
> **Ada Lovelace**

CHAPTER 2

Control Structures

LEARNING OBJECTIVES

At the end of this chapter you should be able to:

O2.A Write MATLAB conditional statements using `if–else` and `switch` statements

O2.B Form complex logical expressions in MATLAB and explain the rules for operator precedence

O2.C Write MATLAB iteration statements using `for` and `while` loops

O2.D Select an appropriate iteration statement for a given programming situation

O2.E Write efficient MATLAB code that uses iteration statements to build arrays

O2.F Make good use of the `break` and `continue` jump statements within MATLAB iteration statements

2.1 INTRODUCTION

In the previous chapter, we introduced the fundamental concepts of computer programming, as well as the basics of the MATLAB software package. In particular, we saw how to create, save and run MATLAB script *m*-files. Scripts allow us to save sequences of commands for future use and can be a very useful time-saving device. MATLAB scripts become more powerful as we learn how to do some simple computer programming. Computer programming not only involves executing a number of commands, but controlling the sequence of execution (or *control flow*) using special programming *constructs*. In this chapter we introduce some of the programming constructs provided by MATLAB.

2.2 CONDITIONAL `IF` STATEMENTS

The normal flow of control in a MATLAB script, and procedural programming in general, is *sequential*, i.e. each program statement is executed in sequence, one after the other. This is illustrated in Fig. 2.1.

MATLAB Programming for Biomedical Engineers and Scientists. DOI: 10.1016/B978-0-12-812203-7.00002-1

FIGURE 2.1 The normal sequential control flow in procedural programming.

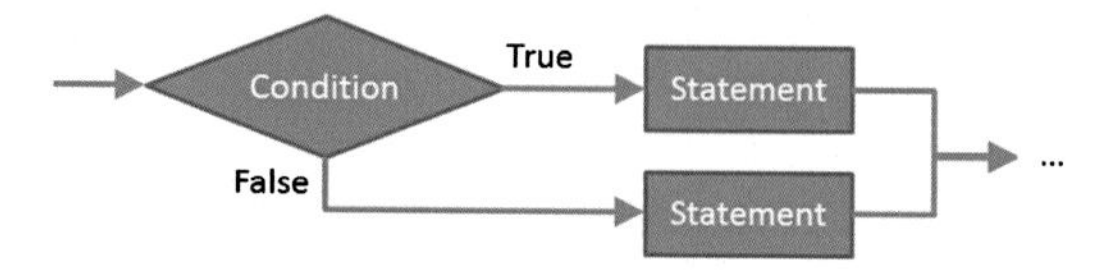

FIGURE 2.2 The control flow of a conditional statement.

For example, to use a medical analogy, a treatment plan might consist of performing a blood test, followed by analyzing the results, and finally administering a particular dose of a drug. Each action is performed one after the other, in sequence. However, with purely sequential control flow we are quite limited in what we can achieve. For instance, we might want to only administer the drug if particular conditions are met in the blood test results. Or, we might want to repeat the blood test after drug delivery to check the patient's response to the drug, and potentially alter the dose accordingly. We might even want to do this many times until the blood test shows a satisfactory response to the drug.

Similarly, in computer programming, if all of our programs featured only sequential control flow they would be limited in their power. To write more complex and powerful programs we need to make use of programming constructs that alter this normal control flow. One type of programming construct is the *conditional* statement. The control flow of a conditional statement is illustrated in Fig. 2.2. In this example there are two possible paths through the program, involving execution of different statements. Which path is taken depends on the result of applying a test *condition*. Returning to our medical analogy, the condition might be based on the results of the blood test, and the two paths of execution might be administering the drug or not.

■ Example 2.1

O2.A

MATLAB `if` statements are one way of achieving a conditional control flow. The use of an `if` statement is illustrated in the example below.

```
a = input('Enter a number:');
if (a >= 0)
    root = sqrt(a);
    disp(['Square root = ' num2str(root)]);
else
    disp(['Number is -ve, there is no square root']);
end
```

■

Try entering these statements and saving them as a MATLAB script *m*-file. Examine the code to make sure that you understand what it does (recall that the `disp` statement displays an array in the command window – in this case an array of characters, or *string*). You can even step through the code using the MATLAB debugger. Pay particular attention to the alternative *paths of execution* that the program can take depending on the result of the comparison `(a >= 0)`. This comparison corresponds to the *condition* box in Fig. 2.2. The two statements after the `if` will be executed only if the condition is true. The statement after the `else` will be executed if the condition is false. After one of these two paths has been taken control will resume immediately after the `end` statement. Note that the `if` and `end` statements are both compulsory and should always come in pairs. The `else` statement is optional and if omitted the program execution path will jump to after the `end` statement if the condition is false.

■ Activity 2.1

In Activity 1.11 you wrote MATLAB code to compute a patient's body mass index (BMI), according to the formula

O2.A

$$\text{BMI} = \text{mass}/\text{height}^2$$

based on a height (in meters) and weight (in kilograms) entered using the keyboard. Modify this code (or write a new version) to subsequently display a text message based on the computed BMI as indicated in the table below.

Range	Message
BMI ≤ 18.5	"Underweight"
18.5 < BMI ≤ 25	"Normal"
25 < BMI ≤ 30	"Overweight"
30 < BMI	"Obese"

■

2.3 COMPARISON/LOGICAL OPERATORS

Example 2.1 introduced the concept of a *condition*. Conditions commonly involve comparisons between expressions involving variables and values. For example, we saw the `>=` comparison operator in the example. A list of common comparison (or *relational*) operators such as `>=` is shown in Table 2.1, along with common *logical* operators for combining the results of different comparisons.

Most of these operators are fairly intuitive. However, note the distinction between `&` and `&&` (and likewise between | and ||). The `&` and | operators perform

Table 2.1 Common relational and logical operators

<table>
<tr><th colspan="2">Relational</th><th colspan="2">Logical</th></tr>
<tr><td>==</td><td>Equals to</td><td>~</td><td>NOT</td></tr>
<tr><td>~=</td><td>Not equals to</td><td>&&</td><td>AND (scalar short-circuit operation)</td></tr>
<tr><td>></td><td>Greater than</td><td>&</td><td>AND (scalar/array operation)</td></tr>
<tr><td><</td><td>Less than</td><td>||</td><td>OR (scalar short-circuit operation)</td></tr>
<tr><td>>=</td><td>Greater than or equal to</td><td>|</td><td>OR (scalar/array operation)</td></tr>
<tr><td><=</td><td>Less than or equal to</td><td></td><td></td></tr>
</table>

AND and OR operations respectively. They will evaluate the expressions on both sides of the operator and return a `true` or `false` value depending on whether both (`&`) or either (`|`) of them evaluated to `true`. These operators will work with either scalar (i.e. single value) logical expressions or array expressions (i.e. that evaluate to an array of `true`/`false` values). The only restriction is that MATLAB must be able to match the expressions on either side of the operator. Either both should be scalars, both should be arrays of the same size, or one should be an array and the other a scalar.

The `&&` and `||` operators perform the same AND and OR operations, but using what is known as a *short-circuiting behavior*. This means that, if the result of the overall AND/OR operation can be determined from the left-hand expression alone, then the right-hand expression will not be evaluated. For example, if the left-hand expression of an AND operation is `false` then the result of the AND will also be `false`, regardless of the value of the right-hand expression. Therefore, the advantage of short-circuiting is that unnecessary operations are not performed. However, *note that `&&` and `||` can only be used with scalar values, not arrays.*

■ Example 2.2

O2.A, O2.B

The use of these relational and logical operators is illustrated in the following code excerpt.

```
...
at_risk = false;
if (gender == 'm') && (calories > 2500)
    at_risk = true;
end
if (gender == 'f') && (calories > 2000)
    at_risk = true;
end
...
```

Here, in both `if` statements, the Boolean `at_risk` variable is set to true only if both comparisons evaluate to true, e.g. if `gender` is equal to 'm' and `calories` is greater than 2500. If either comparison evaluates to false the

Table 2.2 MATLAB operator precedence

Precedence	Operator(s)
1	Brackets ( )
2	Matrix transpose ', power .^, matrix power ^
3	Unary plus +, unary minus −, logical not ~
4	Element-wise multiplication .*, element-wise division ./, matrix multiplication *, matrix division /
5	Addition +, subtraction −
6	Colon operator :
7	Less than <, less than or equal to <=, greater than >, greater than or equal to >=, equal to ==, not equal to ~=
8	Logical and &&
9	Logical or ||

flow of execution will pass to after the corresponding `end` statement. We can use the short-circuiting version of the AND operator as both sides of the operator are scalar (logical) values (i.e. not arrays). ■

Note that there were round brackets around the comparisons in Example 2.2. These are not essential, but they may change the meaning of the condition. For example, suppose the brackets were not included, i.e.

```
...
if gender == 'm' && calories > 2500
...
```

There are three different operators in this condition: ==, && and >. In what order would MATLAB evaluate them? The answer to this question lies in the rules for *operator precedence*. Whenever MATLAB needs to evaluate an expression containing multiple operators it will use the order of precedence shown in Table 2.2. We have come across all of these operators before with a few exceptions. The ' operator, when placed after a matrix, is just a short way of calling the `transpose` function, which we introduced in Section 1.9. The *unary* plus and minus operators refer to a + or − sign being placed before a value or variable to indicate its sign.

Now let's return to our condition without brackets. Of the three operators present (==, && and >), we can see from Table 2.2 that the ones with the highest precedence are == and >. Therefore, these will be evaluated first. As they are equal in precedence they will be evaluated from left to right, i.e. the == first followed by the >. After these evaluations, the condition is simplified to just the && operator with a Boolean value either side of it (whether they are true or false will depend on the values of the `gender` and `calories` variables). Finally, this && operator is evaluated.

In this example, removing the brackets still resulted in the desired behavior. However, this will not always be the case. Furthermore, even if the brackets are not essential it is still a good idea to include them to make our code easier to understand and less prone to errors.

■ Activity 2.2

O2.B

If `a=1`, `b=2` and `c=3`, use your knowledge of the rules of operator precedence to predict the values of the following expressions (i.e. work them out in your head and write down the answers). Verify your predictions by evaluating them using MATLAB.

i `a+b * c`
ii `a ^ b+c`
iii `2 * a == 2 && c - 1 == b`
iv `b + 1:6`

■

2.4 CONDITIONAL `switch` STATEMENTS

Sometimes we may have many `if` statements which all use conditions based on the same variable. It is not incorrect to use `if` statements in such cases, but it can lead to a large number of consecutive `if` statements in our code, making it harder to read and more prone to errors. In this case, it is preferable to use a `switch` statement. The `switch` statement offers an easy way of writing code where the same variable needs to be checked against a number of different values.

■ Example 2.3

O2.A

The following example illustrates the use of a `switch` statement.

```
switch day
    case 1
        day_name = 'Monday';
    case 2
        day_name = 'Tuesday';
    case 3
        day_name = 'Wednesday';
    case 4
        day_name = 'Thursday';
    case 5
        day_name = 'Friday';
    case 6
        day_name = 'Saturday';
    case 7
        day_name = 'Sunday';
    otherwise
        day_name = 'Unknown';
end
```

MATLAB will compare the *switch expression* (in this case, `day`) with each *case expression* in turn (the numbers 1–7). When a comparison evaluates to true, MATLAB executes the corresponding statements and then exits the `switch` statement, i.e. control flow passes to after the `end` statement. The `otherwise` block is optional and executes only when no comparison evaluates to true.

■

Note that the `switch` statement is used only for equality tests – we cannot use it for other types of comparison (e.g. >, <, etc.).

In the above example the switch expression was compared to a single value in each case. It is possible to compare the expression to multiple values by enclosing them within curly brackets and separating them by commas. The corresponding statements are executed if *any* of the values are matched. This is equivalent to an `if` statement with multiple equality tests combined using a || operator.

■ Example 2.4

A `switch` statement with multiple values in each case is illustrated below. *O2.A*

```
switch day
    case {1,2,3,4,5}
        day_name = 'Weekday';
    case {6,7}
        day_name = 'Weekend';
    otherwise
        day_name = 'Unknown';
end
```

■

■ Activity 2.3

Write a MATLAB script *m*-file to read a character from the keyboard, then read a number. Based on the value of the character the program should compute either the *sine* of the number (if the character was an 's'), the *cosine* (if it was a 'c') or the *tangent* (if it was a 't'). If none of these three characters was entered a suitable error message should be displayed. Use a `switch` statement in your program. *O2.A*

(Hint: If you use the `input` command to read a character, by default it will try to 'evaluate' it as a variable. To prevent this, add a second argument 's'. See the MATLAB documentation for `input` for details.) ■

■ Activity 2.4

The normal human heart rate ranges from 60 to 100 beats per minute (BPM) at rest. *Bradycardia* refers to a slow heart rate, defined as below 60 BPM. *O2.A, O2.B*

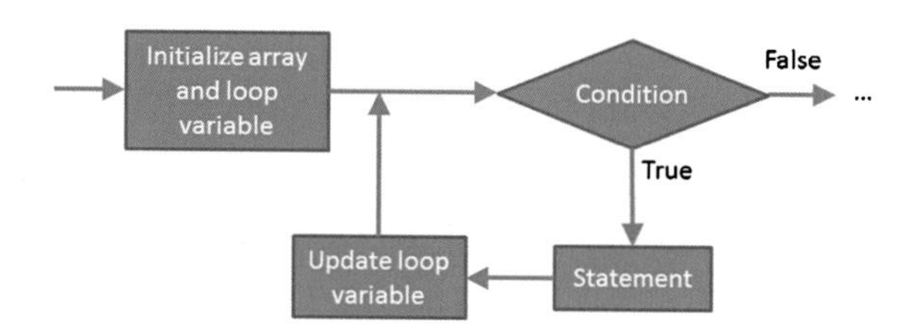

FIGURE 2.3 The control flow of a `for` loop.

Tachycardia refers to a fast heart rate, defined as above 100 BPM. Write a MATLAB script *m*-file to read in a heart rate from the keyboard, and then display the text "Warning, abnormal heart rate" if it is outside of the normal range. ■

■ Activity 2.5

O2.A

Now modify your solution to Activity 2.4 to display one of two warning messages, depending on whether the heart rate indicates bradycardia or tachycardia. If the heart rate is within the normal range the text "Normal heart rate" should be displayed. ■

2.5 ITERATION: `FOR` LOOPS

In addition to conditional statements, the second fundamental type of programming construct that we will consider in this chapter is the *iteration* statement. Iteration statements are intended for use in situations in which one or more operations need to be repeatedly performed a number of times. In computer programming, iteration statements are often known as *loop* statements. Returning to the medical analogy that we introduced in Section 2.2, iteration could involve performing blood tests and modifying the drug dosage repeatedly until certain conditions in the blood test result are met (i.e. the patient has responded satisfactorily).

There are two types of iteration statement provided in MATLAB. In this section, we consider the `for` loop. The control flow of a MATLAB `for` loop is shown in Fig. 2.3. Iteratively, the same statement, or set of statements, is executed a number of times. In Fig. 2.3 the *statement* is continually executed until a certain *condition* becomes false. A *loop variable* will be updated in each iteration to take on a different value from an *array*, which has been initialized at the beginning of the `for` loop. When the statement has been executed once for each element of the array, the condition becomes false.

■ Example 2.5

for loops can be useful when we want to execute some statements a fixed number of times. Consider the following example first, then we will return to the flow chart in Fig. 2.3.

O2.C

```
% ask user for input
n = input('Enter number:');
f = 1;
if (n >= 0)
    for i = 2:n
        f = f * i;
    end
    disp(['The factorial of ' num2str(n) ' is ' num2str(f)]);
else
    disp(['Cannot compute factorial of -ve number']);
end
```

This code reads in a number from the user and computes its factorial. (Note that MATLAB actually has a built-in function to compute factorials: it's called `factorial`.) The code first checks to see if the number is greater than or equal to zero. If not, program execution jumps to after the `else` statement and an error message is displayed. If it is greater than or equal to zero, the program executes the statement `f = f * i` once for each value in the array `2:n`. During each pass through the loop (an *iteration*), the variable `i` takes on the value of the current element of the array. So, for the first iteration `i=2`, for the second iteration `i=3`, and so on, until `i` takes on the value of `n`.

Note how the *assignment* statement inside the `for` loop (`f = f * i`) has the variable `f` appearing on both sides. This might seem a little strange and it is important to understand here that the part of the statement on the right-hand side of the assignment operator (=) is evaluated first using the current value of `f`. The result of multiplying `f` by the current value of the loop variable `i` is found and it is then assigned to `f`. In other words, the value of `f` is updated. After this the `for` loop proceeds to the next iteration.

After the loop has finished all of its iterations (i.e. `i` has taken on all values of the array `2:n` and `f = f * i` has been executed for each one), it exits and program execution continues after the `end` statement. Since `f` was initialized to 1, the effect of iterating through the loop is to multiply 1 by every integer between 2 and `n`. In other words, it computes the factorial of `n`.

Try typing in this code, saving it as a script *m*-file and running it. Read through it carefully to make sure you understand how it works. You can step through it using the debugger to help you. ■

Now let's return to the flow chart in Fig. 2.3. First, the *initialize array and loop variable* box corresponds, in our MATLAB example, to setting up the array `2:n` and initializing `i` to its first element. The *condition* is to test if we have come to the end of the array. When we reach the end, the condition becomes false and the loop will exit. The *statement* corresponds to `f = f * i`. The *update loop variable* box will modify `i` so that it takes on the value of the next element in the array.

■ Activity 2.6

O2.C

Write a MATLAB script *m*-file that uses a `for` loop to compute the value of the expression:

$$\sum_{a=1}^{5} a^3 + a^2 + a$$

■

■ Activity 2.7

O2.A, O2.C

In a *phase I* clinical trial for a new drug one of the main goals is to determine the recommended dose of the drug for phase II of the trial. Typically, the drug is tested on a small cohort of patients with a low dose, and efficacy and toxicity are estimated. Then, the dose is *escalated* (i.e. increased) and efficacy/toxicity are estimated on a new cohort. This process is continued until toxicity becomes too high. The dose/toxicity results inform the choice of drug dose for phase II of the trial [2].

The increasing dose levels used in a phase I trial are known as the *dose escalation pattern*. A common way of calculating the dose escalation pattern is to increase the dose by 67% after the first cohort, by 50% after the second, by 40% after the third and by 33% after each subsequent cohort.

Write a MATLAB script *m*-file to determine the dose escalation pattern for a new cytotoxic drug that is being tested for cancer treatment. The starting dose should be 0.2 mg/kg. The script should determine the first ten drug doses in the escalation pattern. These should be (in mg/kg):

```
0.2
0.334
0.501
0.7014
0.932862
1.240706
1.65014
2.194686
2.918932
3.882179
```

■

Generally, `for` loops are appropriate when it is known, before starting, exactly how many times the loop should be executed. For instance, in Example 2.5 the

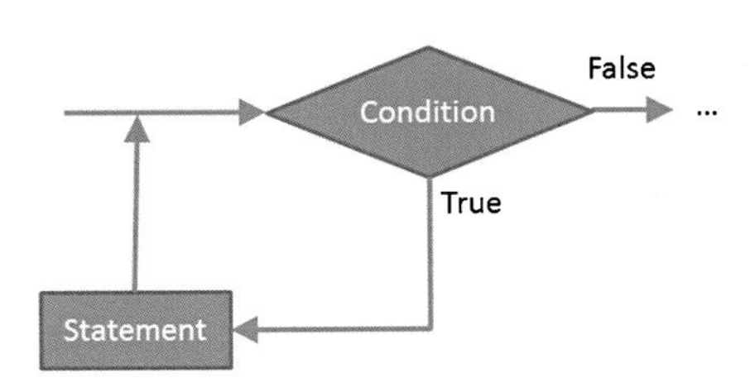

FIGURE 2.4 The control flow of a `while` loop.

value of `n` was entered by the user so the number of loop iterations could be determined from this value. If it is not possible to know how many times the loop should be executed, an alternative iteration construct is necessary, as will be described below.

2.6 ITERATION: `WHILE` LOOPS

The second type of iteration statement available in MATLAB is the `while` loop. The control flow of a `while` loop is shown in Fig. 2.4. First, a condition is tested. If this condition evaluates to false then control immediately passes beyond the `while` statement. If it evaluates to true then a statement (or sequence of statements) is executed and the condition is retested. This *iteration* continues until the condition evaluates to false.

Therefore, a `while` loop will execute continually until some condition is no longer met. This type of behavior can be useful when continued loop execution depends upon some computation or input that cannot be known before the loop starts.

■ Example 2.6

Consider the following simple example.

O2.C

```
i=randi(10) -1; % random integer between 0 and 9
guess = -1;
while (guess ~= i)
    guess = input('Guess a number:');
    if (guess == i)
        disp('Correct!');
    else
        disp('Wrong, try again ...');
    end
end
```

This program first generates a random integer between 0 and 9. The MATLAB `randi` statement, used in this way, generates a random integer from a uniform distribution between 1 and its argument (in this case, 10), so by subtracting 1 from this we can get a random integer between 0 and 9. The following `while` loop will iterate whilst the condition `guess ~= i` is true. In other words, the loop will execute so long as `guess` is not equal to the random integer `i`. Since `guess` is initialized to −1 before the loop starts, the loop will always execute at least once. Inside the loop, a number is read from the user, and an `if` statement is used to display an appropriate message depending upon whether the guessed number is correct. The program will continue asking for guesses until the correct answer is entered. ■

Generally, `while` loops are useful when it is not known in advance how many times the loop should be executed. Both `for` loops and `while` loops can be very useful, but in different types of situation. In time you will learn to recognize when to use each of these types of iteration programming construct.

■ Activity 2.8

O2.C

Write a MATLAB script *m*-file that continually displays random real numbers between 0 and 1, each time asking the user if they want to continue. If a 'y' is entered the program should display another random number, if not it should terminate. Use a `while` loop in your implementation. ■

■ Activity 2.9

O2.C, O2.D

Modify your solution to Activity 2.8 so that rather than displaying a single random value, it asks the user *how many* random values they would like. If they enter a number greater than zero the program displays that many random values. Otherwise it should repeat the request for the number of random values. The program should still terminate when any character other than a 'y' is entered when the user is asked if they want to continue. ■

2.7 A NOTE ABOUT EFFICIENCY

■ Example 2.7

O2.C, O2.E

There are many programming situations in which an array of values needs to be built up inside a loop. For example, the following piece of code uses a `for` loop to build up an array of 1,000,000 random real numbers between 0 and 1.

```
clear
tic
nRands = 1000000;
for i=1:nRands
    rand_array(i) = rand(1);
end
toc
```

The code also uses some built-in MATLAB functions (`tic` and `toc`) to measure how long it takes to build the array. However, the code is not particularly efficient. ■

The reason for the inefficiency of this code is that in MATLAB, arrays are *dynamically* allocated. This means that memory is set aside (or *allocated*) for the array based on the highest indexed element that we have created so far. If we attempt to place a value into the array at an index that is greater than its current size, MATLAB will recognize that we need a larger array and it will automatically resize the original array to allow the assignment. This is a nice feature that makes programming easier at times. However, it comes with a cost: our program may not execute as quickly. This is because sometimes MATLAB may need to allocate a new, larger block of memory and copy all the values from the block where the array is currently stored. In summary, if we know (or have a good guess at) how big the array needs to be in the first place, we can "pre-allocate" it and our program is likely to be more efficient.

■ Example 2.8

The same piece of code from the previous example is shown below, now with a pre-allocation of the array.

O2.C, O2.E

```
clear
tic
nRands = 1000000;
rand_array = zeros(1,nRands);
for i=1:nRands
    rand_array(i) = rand(1);
end
toc
```

Here, the `zeros` function is used to pre-allocate a large array of zeros. Therefore, MATLAB knows straight away how large a block of memory to set aside and will never have to allocate a new block and copy data across. Thus, the code should run quicker. You can verify this yourself by typing in the two versions of the code and running them. The code should measure how long it took to execute each version and display this information. ■

We will return to the important topic of code efficiency in Chapter 9.

2.8 BREAK AND CONTINUE

Another way of altering the control flow of a program is to use jump statements. The effect of a jump statement is to unconditionally transfer control to another part of the program. MATLAB provides two different jump statements: `break` and `continue`. Both can only be used inside `for` or `while` loops.

The effect of a `break` statement is to transfer control to the statement immediately following the enclosing control structure.

■ Example 2.9

O2.A, O2.C, O2.F

As an example, consider the following code.

```
total = 0;
while (true)
    n = input('Enter number: ');
    if (n < 0)
        disp('Finished!');
        break;
    end
    total = total + n;
end
disp(['Total = ' num2str(total)]);
```

This piece of code reads in a sequence of numbers from the user. When the user types a negative number the loop is terminated by the `break` statement. Otherwise, the current number is added to the `total` variable. When the loop terminates (i.e. a negative number is entered), the value of the `total` variable, which is the sum of all of the numbers entered, is displayed. ■

The `continue` statement is similar to the `break` statement, but instead of transferring control to the statement following the enclosing control structure, it only terminates the current iteration of the loop. Program execution resumes with the next iteration of the loop.

■ Example 2.10

O2.A, O2.C, O2.F

For example, examine the following piece of code.

```
total = 0;
for i = 1:10
    n = input('Enter number: ');
    if (n < 0)
        disp('Ignoring!');
        continue;
    end
    total = total + n;
end
disp(['Total = ' num2str(total)]);
```

A sequence of numbers is again read in and summed. However, this time there is a limit of 10 numbers, and negative numbers are ignored. If a negative number is entered, the `continue` statement causes execution to resume with the next iteration of the `for` loop. ■

■ Activity 2.10

O2.A, O2.C, O2.F

The listing below shows a MATLAB program that reads 20 numbers from the user and displays the sum of their square roots.

```
sum = 0;
for x=1:20
    n = input('Enter a number:');
    sum = sum + sqrt(n);
end
disp(['Sum of square roots = ' num2str(sum)]);
```

Enter this script into a MATLAB *m*-file and modify it so that:

- negative numbers are ignored, and
- entering the number 0 causes the program to display the sum so far and then terminate. ■

2.9 NESTING CONTROL STRUCTURES

Any of the control structure statements we have covered in this chapter can be *nested*. This simply means putting one statement inside another one. We have already seen nested statements in some of the examples and exercises of this chapter.

■ Example 2.11

O2.A, O2.C, O2.D

As another example, examine the following piece of code and see if you can work out what it does.

```
n = input('Enter number: ');
while (n >= 0)
    f = 1;
    for i = 2:n
        f = f * i;
    end
    disp(f);
    n = input('Enter number: ');
end
```

Here, we have a `for` loop nested inside a `while` loop. The `while` loop reads in a sequence of numbers from the user, terminated by a negative number. For each positive number entered, the `for` loop computes its factorial, which is then displayed. ■

2.10 SUMMARY

Control structures allow us to alter the natural sequential flow of execution of program statements. Two fundamental types of control structure are *conditional* statements and *iteration* statements. MATLAB implements conditional control flow using the `if–else–end` and `switch` statements. Iterative control flow can be implemented with `for` loops and `while` loops. All of these control structures can be *nested* inside each other to produce more complex forms of control flow. The `break` and `continue` statements allow 'jumping' of control from inside loop statements.

2.11 FURTHER RESOURCES

- MATLAB documentation on control flow statements: http://www.mathworks.co.uk/help/matlab/control-flow.html.

EXERCISES

■ Exercise 2.1

O2.A, O2.B

Cardiac resynchronization therapy (CRT) is a treatment for heart failure that involves implantation of a pacemaker to control the beating of the heart. A number of indicators are used to assess if a patient is suitable for CRT. These can include:

- New York Heart Association (NYHA) class 3 or 4 – this is a number representing the severity of symptoms, ranging from 1 (mild) to 4 (severe);
- 6 minute walk distance (6MWD) less than 225 meters; and
- Left ventricular ejection fraction (EF) less than 35%.

Write a MATLAB script *m*-file containing a single `if` statement that decides if a patient is suitable for CRT, based on the values of three numerical variables: `nyha`, `sixmwd` and `ef`, which represent the indicators listed above. All three conditions must be met for a patient to be considered suitable. You can test your program by assigning values to these variables from the test data shown in the table below.

Patient	NYHA class	6MWD	EF	Suitable for CRT?
1	3	250	30	No
2	4	170	25	Yes
3	2	210	40	No
4	3	200	33	Yes

■

■ Exercise 2.2

O2.A, O2.B

Not all patients respond positively to CRT treatment. Common indicators of success include:

- Decrease in NYHA class of at least 1;
- Increase in 6MWD of at least 10% (e.g. from 200 m to 220 m); and
- Increase in EF of at least 10% (e.g. from 30% to 40%).

The table below shows post-CRT data for patients 2 and 4 from Exercise 2.1 (who were considered suitable for CRT). Write a MATLAB script *m*-file that determines if a patient has responded positively to CRT treatment. All three conditions must be met for a patient to be considered a positive responder. The program should use a single `if` statement which tests conditions based on the values of six numerical variables: `nyha_pre`, `sixmwd_pre`, `ef_pre`, `nyha_post`, `sixmwd_post` and `ef_post`, which represent the pre- and post-treatment indicators.

Patient	NYHA class	6MWD	EF	Responder?
2	3	220	30	No
4	2	230	45	Yes

■

■ Exercise 2.3

O2.A

Write a MATLAB script *m*-file that reads in two numbers from the keyboard followed by a character. Depending on the value of the character, the program should output the result of the corresponding arithmetic operation ('+', '−', '∗' or '/'). If none of these characters was entered a suitable error message should be displayed. Use a `switch` statement in your program. ■

■ Exercise 2.4

O2.A, O2.B

The figure below shows how blood pressure can be classified based on the diastolic and systolic pressures. Write a MATLAB script *m*-file to display a message indicating the classification based on the values of two variables representing the diastolic and systolic pressures. The two blood pressure values should be read in from the keyboard.

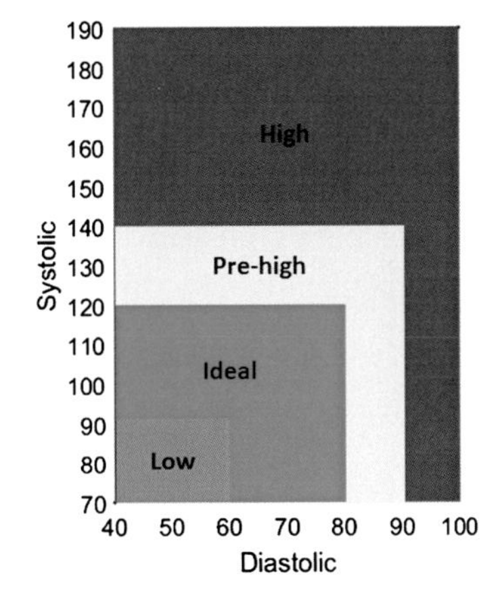

■

■ Exercise 2.5

O2.A

Write a `switch` statement that tests the value of a character variable called `grade`. Depending on the value of `grade` a message should be displayed as shown in the table below. If none of the listed values are matched the text "Unknown grade" should be displayed.

Grade	Message
'A' or 'a'	"Excellent"
'B' or 'b'	"Good"
'C' or 'c'	"OK"
'D' or 'd'	"Below average"
'F' or 'f'	"Fail"

■

■ Exercise 2.6

O2.C, O2.D

Write a MATLAB script *m*-file to read in an integer N from the keyboard and then use an appropriate iteration programming construct to compute the value of the expression:

$\sum_{n=1}^{N} \frac{n+1}{\sqrt{n}} + n^2$ ■

■ Exercise 2.7

O2.C, O2.D, O2.E

Write a MATLAB script *m*-file to compute the value of the expression

$\sin(x) + \frac{1}{2}\cos(x)$ ■

for values of x between 0 and 2π in steps of 0.0001, and produce a plot of the results.

Write two separate MATLAB script *m*-files to implement this task: one that produces the result using MATLAB array processing operations as we learned about in Chapter 1; and one that builds up the array result element-by-element using an iteration programming construct. Add code to time your two implementations. Which is quicker and why?

■ Exercise 2.8

O2.C, O2.D

In Activity 1.8 we introduced the field of gait analysis. The file *LAnkle_tracking.mat* contains data acquired using an optical tracking system from the same patient mentioned in Activity 1.8, whilst they were walking. The file contains a single variable called `LAnkle`, which is a matrix containing four columns. The first column represents timings in seconds, and the next three columns represent the x, y and z coordinates of a marker attached to the patient's left ankle.

In this exercise you will write code to automatically identify a particular gait event: the *foot strike* of the left leg, which refers to the instant that the left heel touches the ground. The foot strike event can be determined by looking for *troughs* in the series of z coordinates of the ankle marker. The z coordinates represent the *inferior–superior* (i.e. foot–head) motion of the marker.

Write a MATLAB script *m*-file that loads in the *LAnkle_tracking.mat* file and first produces a plot of the z-coordinates against time. The script should then search the list of z-coordinates and report any troughs. To begin with, consider a trough to be any value that is less than both the preceding and following values.

Test your code. By looking at the plot and the identified foot strike events, do you think it is working correctly? What can you do to rectify any problems? ■

■ Exercise 2.9

O2.C, O2.D

The Taylor series for the exponential function e^x is:

$$1 + \frac{x^1}{1!} + \frac{x^2}{2!} + \frac{x^3}{3!} \cdots$$

Write a MATLAB script *m*-file to read a number x from the keyboard and compute its exponential using the Taylor series. Also compute the exponential using the built-in MATLAB function `exp`. Run your program several times with different numbers of terms in the Taylor series to see the effect this has on the closeness of the approximation. ■

■ Exercise 2.10

O2.A, O2.C, O2.D

Modify your solution to Exercise 2.9 so that, as well as computing and displaying the result of the Taylor series expansion, it also computes and displays the number of terms required for the approximation to be accurate to within 0.01. ■

■ Exercise 2.11

O2.A, O2.C, O2.D, O2.F

Modify your solution to Activity 2.3 so that, rather than reading a single character/number, the program continually reads character/number pairs and displays the trigonometry result. The program should terminate when the character 'x' is entered. In this case, no number should be read and the program should exit immediately. If a number outside the range $[-100, 100]$ is entered, the program should not display the trigonometry result and should ask for another character/number pair to be entered. ■

■ Exercise 2.12

O2.C, O2.D

In Activity 1.6 we introduced the concept of the Injury Severity Score (ISS). The ISS is a medical score to assess trauma severity and is calculated as follows. Each injury is assigned an Abbreviated Injury Scale (AIS) score in the range [0–5] which is allocated to one of six body regions (head, face, chest, abdomen, extremities and external). Only the highest AIS score in each body region is used. The three most severely injured body regions (i.e. with the highest scores) have their AIS scores squared and added together to produce the ISS. Therefore, the value of the ISS is always between 0 and 75. Write a MATLAB script *m*-file to compute the ISS given an array of 6 AIS scores, which represent the most severe injuries to each of the six body regions. Test your code using the array `[3 0 4 5 3 0]`, for which the ISS should be 50. ■

■ Exercise 2.13

O2.A, O2.C, O2.D

In Exercise 1.6 you wrote MATLAB code to check if an integer was divisible by 9 by summing its digits: if the digits add up to 9 the number is divisible by 9. For integers up to 90 only a single application of the rule is required. For larger integers, multiple applications may be required. For example, for 99, the digit sum is 18. Then, because 18 still has multiple digits we sum them again to get 9, confirming that 99 is divisible by 9 (after two steps). In general, for a given starting number, you should repeatedly sum the digits until the result is a single digit number.

Write a MATLAB script *m*-file to read in a single number from the keyboard, and use the rule to determine if it is divisible by 9. An appropriate message should be displayed depending on the result. You can assume that the number will always be less than 1000. ■

FAMOUS COMPUTER PROGRAMMER: ALAN TURING

Alan Turing was an English mathematician and computer scientist who was born in 1912. At school he was criticized by his teachers for not focusing on his work. His headmaster wrote of him: "If he is to stay at Public School, he must aim at becoming educated. If he is to be solely a Scientific Specialist, he is wasting his time at a Public School."

Despite these criticisms, Turing went on to study mathematics at King's College Cambridge. He became interested in algorithms and in 1936 he published a now-famous paper in which he introduced the idea of a 'Turing machine'. This was similar in concept to Charles Babbage's Analytical Engine (see Chapter 1's Famous Computer Programmer), but a fundamental contribution of Turing's proposal was that the data and instructions would be stored in the same computer memory. This makes program development much easier and is the way in which today's digital computers are designed.

In 1939, after the outbreak of the Second World War, Turing started to work at the UK Government Code and Cypher School at Bletchley Park. He was brilliant at cracking codes and developing electro-mechanical machines to assist him in code decyphering. In 1945, he was awarded an OBE for his contributions to the war effort (he kept his OBE in his tool box). After the war Turing performed much of the pioneering work in developing the early digital computers, and designing algorithms and writing code for them.

However, several years later Turing's life took a more tragic turn. In 1952 he was arrested for violation of Britain's laws against homosexuality. At the trial, he offered no defense except that he saw no wrong in what he had done. Alan Turing died of cyanide poisoning in 1954. Cyanide was found on a half eaten apple beside his body. An inquest returned a verdict of suicide but his mother always maintained that it had been an accident.

In September 2009 the UK Prime Minister made an official apology for "the appalling way [Turing] was treated." In December 2013, Alan Turing was granted a posthumous royal pardon. In granting the pardon, the UK Justice Minister described Turing as "an exceptional man with a brilliant mind".

> *"How can one expect a machine to do all this multitudinous variety of things? The answer is that we should consider the machine as doing something quite simple, namely carrying out orders given to it in a standard form which it is able to understand."*
>
> **Alan Turing, 1946**

CHAPTER 3

Functions

LEARNING OBJECTIVES

At the end of this chapter you should be able to:

- *O.3A* Write and run a MATLAB *m*-file that contains a function which can take arguments and return output
- *O.3B* Write and run a MATLAB script *m*-file that just contains a sequence of commands
- *O.3C* Explain the meaning of the scope of a variable in a function
- *O.3D* Include checks for errors or warnings in your own functions
- *O.3E* Set the MATLAB path so that functions you write can be called from any location in the file system
- *O.3F* Write recursive functions in MATLAB

3.1 INTRODUCTION

In this chapter, we will consider the programming construct known as the 'function'. We have already seen how MATLAB provides a large number of built-in functions for our use, so you should be aware that *a function is a piece of code that can take some arguments as input and carry out some tasks that can produce a result*. Functions can be particularly useful if we want to perform exactly the same operation many times with different input values. By learning to write our own functions, we can take advantage of a lot of the power available in MATLAB (or indeed in many other programming languages).

3.2 FUNCTIONS

MATLAB has its own built-in functions such as `sin`, `*`, `exp`, etc. but it also allows us to define our own functions. We can save any function we write in an *m*-file in the same way as we can save any set of commands to form a script. In both cases this helps us to share our work with others.

MATLAB Programming for Biomedical Engineers and Scientists. DOI: 10.1016/B978-0-12-812203-7.00003-3

So it is important to note that there *two kinds of* m-*file*. There are *function m*-files and *script m*-files. We have already seen script *m*-files in Section 1.10. Later, we will discuss them and how they differ from function *m*-files in Section 3.4.

The following examples illustrate some functions.

■ Example 3.1

O2.A

Here is a very simple (and quite silly) function.

```
function out = add_two(in)
% Usage:
%  out = add_two(in)
%  in  : Argument passed in
%  out : Output, result of adding two to input.

out = in + 2;
end
```

This very simple function has a single input argument, `in`, and a single output argument, `out`, which is the result of adding two to the input.

Note how the code begins with the `function` keyword. This is then followed by the name of the output argument `out`, then an equals sign, then the name of the function `add_two`. The first line is then ended by the input argument, `in`, inside a pair of brackets. This first line is known as the function's *prototype*. It is also sometimes described as the function *definition* or *signature*.

After the first line, some comments describe how the function should be used and then the 'work' of the function is actually carried out in the line `out=in+2;`. We indicate the end of the function with the `end` keyword.

The code for a function can be saved in a file. ■

■ Activity 3.1

O3.A

Use the *editor* window (see Fig. 1.1) to type the code for the function from Example 3.1 into a new file. Then save the file with the name *add_two.m*. This kind of file is called a *function m*-file.

Once the file containing the function is saved, go to the command window and call the function as follows:

```
>> a = 5
a =

     5
>> b = add_two(a)
b =

     7
```

In the above, we assign a value of 5 to the variable with the name `a` and then pass it to the function as an input argument. The return value from the function is assigned to the variable named `b`.

Experiment with different input arguments and assign the return value to variables with different names. ■

Note that a function can have any number of input arguments (including none) and any number of outputs (including none).

■ Example 3.2

The next function takes two input arguments. It does not return any output arguments. It simply prints the sum of the inputs to the command window when it runs. *O3.A*

```
function report_sum(x, y)
% Usage:
%   report_sum(x, y)
%   x, y : numbers to be added

fprintf('The sum is %u\n', x + y)
end
```

Here, we have used the built-in `fprintf` function to print output to the screen. We will discuss this function in more detail in Section 6.5. ■

■ Activity 3.2

Type the code from Example 3.2 in the editor window and save it in a function *m*-file with the name *report_sum.m*. *O3.A*

Once the file containing the function is saved, go to the command window and call the function to make sure it works, for example as follows:

```
>> report_sum(3, 25)
The sum is 28
```

Note how there was no need to assign any return value to a variable, because the function `report_sum` did not have any output arguments. ■

■ Example 3.3

The next function takes no inputs and gives no return value! It simply prints one of the words 'low' or 'high' to the screen at random. *O3.A*

```
function say_low_or_high()
% Usage:
%   say_low_or_high

a = rand();
if a < 0.5
```

```
  fprintf( 'low\n' )
else
  fprintf( 'high\n' )
end

end
```

Here, we have again used the built-in MATLAB function `fprintf` to print output to the screen. We have also used the built-in function `rand()` to generate a random number in the interval [0, 1] (type `help rand` at the command window to get an idea of how it works). ■

■ Example 3.4

O3.A

This example is slightly more complex than the previous ones.

```
function [l,m] = compare_to_mean(x)
% Usage:
%   [l,m] = compare_to_mean(x)
% Input:
%     x - A list of numbers
% Outputs:
%     l - Number of values in x less than mean
%     m - Number of values in x greater than mean

m = mean(x);

less_array = find(x < m);
l = length(less_array);

more_array = find(x > m);
m = length(more_array);

end
```

We will use and discuss this function in more detail in the next few pages. ■

■ Activity 3.3

O3.A

Create a new file in the editor window and type in the code for the function `compare_to_mean`. Save it with the file name *compare_to_mean.m*.

Now switch to the command window and call your function by typing the code below. First, we create an array with some numbers of our choosing, e.g.

```
>> a = [1 2 4 9 16 25 36 49 64 81 100];
```

Now, we check the function works by passing our array as input:

```
>> [countLess, countMore] = compare_to_mean(a)
countLess =
     6
countMore =
     5
```

We find that six elements of the array were below the mean value and five were above the mean value in this example. Check for other arrays of your own. ■

Note how the call to the function in Activity 3.3 captures *both* return values and assigns them to the two variables called `countLess` and `countMore`. Inside the function code, these are the values of the output arguments that are returned (with the names `l` and `m`).

While the function `compare_to_mean` will always return two output arguments, when we call the function, we do not have to assign all of them to variables in the command window call. For example, the following call

```
x = compare_to_mean(a)
```

will only capture the value of the *first* of the return values (`l` in the original function), assigning it to a command window variable called `x`. In this case, we collect fewer values from the function than it returns and the second return value is simply lost or discarded.

However, if we try to collect *more* values than the function can provide, we will obtain an error:

```
>> [a, b, c] = compare_to_mean(a)

 Error using compare_to_mean
 Too many output arguments.
```

Similarly, we can encounter errors if we do not pass the correct number of input arguments to a function. Using the function `report_sum` from Example 3.2, MATLAB will give an error message in the following examples

```
    >> report_sum(3,4,5)

 Error using report_sum
 Too many input arguments.


    >> report_sum(3)

 Not enough input arguments.
 Error in report_sum (line 6)
 fprintf('The sum is %u\n', x + y)
```

We give too many inputs in the first case and not enough inputs in the second.

As we have seen in all of these examples, we can create our own functions and we can call them from the command window in the same way that we can call built-in MATLAB functions.

The function in Example 3.4 takes an array as its argument and returns the counts of the array elements that are less than or greater than the mean value of the array elements. Therefore, we can assign the return values of the function to *two* variables, enclosed by square brackets.

Let's look more closely at the `compare_to_mean` function code. The first line is the function *prototype* (a.k.a. *definition* or *signature*). The prototype line looks like this:

```
function [l,m] = compare_to_mean(x)
```

This important line specifies

- That the code represents a function, using the `function` keyword.
- The name of the function: `compare_to_mean`.
- The input argument(s) that the function expects: in this case just one input, `x`.
- The output values that it will return (`l` and `m`).

We do *not* need to state the *types* of the values (e.g. number, character, array, see Section 1.5). This is a feature of programming using MATLAB and is in contrast to many other programming languages. The reason for this difference is that MATLAB programming uses *dynamic typing*, i.e. we can assign values to variables without specifying their type, and the type of a variable can also change during the running of a MATLAB program.

The next set of lines in Example 3.4 are all comments, as they start with the `%` symbol:

```
% Usage:
% ...
%      m - Number of values in x greater than mean
```

As pointed out in Section 1.11, it is good practice to include comments like this to help other people (and ourselves) to understand and make use of our functions. The comments at the start of a function are used by MATLAB when the built-in `help` command is called for our function. If we type `help compare_to_mean` in the command window, the comments at the top of the function are printed to the screen. As noted in Section 1.3, this works for all built-in MATLAB functions, e.g. try typing `help sin` in the command window.

After the comments at the start of `compare_to_mean`, a series of steps are carried out to do the work of the function:

- The mean value of the input array, x, is computed and assigned to the variable m.
- The built-in find command is used to find the elements of the array that are less than m. Note that the expression x < m will return an array containing 1s (true) where the original element was less than m and 0s (false) where they weren't. The find command will return a new array containing the indices (i.e. the array element numbers) of the non-zero elements of this array.
- Therefore, the subsequent call to obtain the length of the array returned by the find command will give the number of the original array elements that were less than the mean. This is assigned to the variable l, one of the return variables specified in the function prototype.
- An almost identical pair of steps are then taken to assign a value to the other return variable, m.

Read through this code carefully and make sure you understand how it works.

3.3 CHECKING FOR ERRORS

We often need to check for errors during the running of a function and, if we find one, we can return from it before it completes all its tasks. One common check is to ensure that the user has called the function with the correct arguments. For instance, in Example 3.4 the user might give an empty array as the input in the command window, i.e.

```
>> a = [];
>> [lo, hi] = compare_to_mean(a);
```

In this case the function should return before attempting any calculation. This can be achieved using the built-in error function. Example 3.4 can be adjusted slightly to address this by using the built-in isempty function as shown below:

```
function [l,m] = compare_to_mean(x)
% [Usage and help section goes here ....]

if isempty(x)
  error('compare_to_mean: %s', 'The input array is empty.');
end

% [... function continues as before]
```

The error call has two arguments: the first is a string which contains a *formatting character* %s. This formatting character is substituted with the second argument to make a single string which is used as the message displayed by the error function.

When the if condition is true and the user has given an empty array, the full formatted message 'compare_to_mean: The input array is empty.' is printed in the command window and the function exits immediately without giving any return value. It is important that the error message contains some text to say which function was being called when the error occurred, especially in complex code when lots of functions are being called.

Other format strings are possible when using the `error` command. For example `%u` can be substituted with a subsequent integer argument. For more information on the `error` function use `doc` or `help` at the command window. We will return to the topic of format strings in Section 6.4.

■ Activity 3.4

O3.D

Use the code for the `compare_to_mean` function in Example 3.4 and include a check to see if the correct number of input arguments has been given. There should be exactly one argument and the function should give an `error` call if not.

(Hint: Use the built-in function `nargin` to check how many arguments have been passed to the function. Type `doc nargin` or `help nargin` at the command window to see how it works.) ■

3.4 FUNCTION *m*-FILES AND SCRIPT *m*-FILES

Recall that we described script *m*-files in Section 1.10. To recap, a script *m*-file simply contains a sequence of MATLAB commands. It is different from a function *m*-file because it does not take any input argument(s) nor return any value(s). Also, it does *not* have a function prototype line at the top of the file.

The following table lists the main differences between function *m*-files and script *m*-files

Script *m*-file	Function *m*-file
A simple list of commands	Commands enclosed by *keywords*
No `function` keyword at start	`function` keyword at start of file
No input arguments	Prototype line at start lists input arguments
No output arguments	Prototype line lists output arguments
No keyword when finished	`end` keyword to indicate we have reached the end of the function.

Note that the use of the `end` keyword at the end of a function *m*-file is not compulsory but it is considered to be good practice.

■ Example 3.5

O3.B

This example is a script *m*-file. It is just a list of commands, and no function keyword or prototype is used.

```
disp('Hello world')

total = 0;

for i = 1:1000
    total = total + i;
end

fprintf('Sum of first 1000 integers: %u\n' , total)
```

■

■ Activity 3.5

Type the commands from Example 3.5 into a script *m*-file and save it with the name *run_some_commands.m*

O3.B

Run the script at the command window by simply calling the name of the file (without the `.m` suffix). Check that it gives the correct output as shown below.

```
>> run_some_commands
Hello world
The sum of the first 1000 integers is 500500
```

What would you need to do to make the script produce a different output, such as the following?

```
>> run_some_commands
Hi Universe!
The sum of the first 10 integers is 55
```

■

Important: The script *run_some_commands.m* will *always do the same thing* unless we modify it. If we want to sum the first 100 or 1,000,000 numbers, we need to edit the file and change it to give us what we want. This is one reason why it is generally better to write a function that can take an input argument.

■ Example 3.6

We can create a function to do something similar but with input and output arguments as follows:

O3.A

```
function total = sum_function(N)

total = 0

for i = 1:N
    total = total + i;
end

end
```

■

In this case, the function does not specify the number `N`, it is an input argument that the user can pass in and any integer will suffice.

■ Activity 3.6

O3.B

Use the editor window to type in the code for the function `sum_function` and save it in a function *m*-file called *sum_function.m*.

We can call the function in a similar way to the script but now we must pass an input argument using brackets. Check that the function returns the correct output for the sum of the first 1000 integers, i.e.

```
>> value = sum_function(1000)

value =

  500500
```

Show how you can find the sum of the first 20, 50, and 300 integers by simply modifying how you call the function at the command window (you do not need to adjust the function's *m*-file!). ■

3.5 A FUNCTION *m*-FILE CAN CONTAIN MORE THAN ONE FUNCTION

If we are working on an *m*-file that contains a function that gets long and complicated, we can break it up into a 'main' function and pass on some of its work to be carried out in sub-functions or helper functions. The sub-functions can appear later within the same file after the 'main' function. The following example illustrates how to do this.

■ Example 3.7

O3.A

The code below represents three functions, and all of them can be saved in a single file called *get_basic_stats.m*. The 'main' function is called `get_basic_stats` and there are two helper functions that follow it.

```
function [xMean, xSD] = get_basic_stats(x)
% Usage :
%   [xMean, xSD] = get_basic_stats(x)
%
% Return the basic stats for the array x
% Input: x, array of numbers
% Outputs: mean, SD of x.

xMean = get_mean(x);
xSD   = get_SD(x);

end
```

```
% Helper functions below:
function m = get_mean(x)
m = mean(x);
end

function s = get_SD(x)
s = std(x);
end
```

■

This example illustrates the following rules that must be followed:

- The main function appears first (`get_basic_stats`).
- The name of the main function and the *m*-file must match.
- The helper functions are typed in after the `end` of the main function. (In the above example the helper functions are `get_mean` and `get_SD`.)
- It is the main function that calls the helper functions.

Note that these helper functions *can only be called by the main function in this file*. They cannot be called, for example, from the command window or from a function in another *m*-file.

Good advice: When typing multiple functions into a single *m*-file, make sure to `end` one function before starting another. Otherwise, there can be an error or unpredictable behavior. In other words the structure of a file with more than one function in it should be:

```
function val = f1(input1)

% ... f1 code here

end % for f1

function val = f2(input2)

% ... f2 code here

end % for f2

% etc..
```

It is even better to put a comment line in between functions to make the code readable, as in the following

```
function val = f1(input1)

% ... f1 code here

end % for f1

%%%%%%%%%%%%%%%%%%%%
```

```
function val = f2(input2)

% ... f2 code here

end % for f2
```

(It is actually possible to nest one function inside another but you are *strongly* advised not to do this. It is rarely used and only in special circumstances.)

3.6 A SCRIPT *m*-FILE CANNOT ALSO INCLUDE FUNCTIONS

As we saw above, a script *m*-file is just a list of commands. The following is a modified version of the code in Example 3.5 and it shows what *not* to do.

```
% Start of script m-file called 'bad.m'
disp('Hello world')

total = find_sum_illegal(1000);

fprintf('Sum of first 1000 integers: %u\n' , total)

function tot = find_sum_illegal(N)
tot = 0;
for i = 1:1000
    tot = tot + i;
end

end
% End of script m-file called 'bad.m'
```

This is illegal because the file has started as a script and later on it introduces a function. MATLAB will not allow this to happen. To make the above work, we would need to separate the script part and function part and put them in separate files: a script *m*-file and a function *m*-file.

■ Example 3.8

O3.A, O3.B

This example shows the correct separation of the code above.

Script *m*-file:

```
% In a script m-file run_some_commands_2.m
disp('Hello world')

total = find_sum_legal(1000);

fprintf('Sum of first 1000 integers: %u\n' , total)
```

Function *m*-file:

```
% In a function m-file find_sum_legal.m
function tot = find_sum_legal(N)
tot = 0;
for i = 1:N
    tot = tot + i;
end

end
```

■

3.7 *m*-FILES AND THE MATLAB SEARCH PATH

When using *m*-files to save functions, we may need to save them in different locations in the file system. This relates to the set of directories in which MATLAB will search for *m*-files. Collectively, the set of all locations where MATLAB searches is called the *path* and we can inspect the current set of locations by typing the built-in function `path` in the command window.

Recall that the current working directory is shown near the top of the MATLAB environment (see Fig. 1.1). If a function *m*-file that we have written is in the current working directory, then we can call it from the command window directly.

If, on the other hand, the function that we want to call has its *m*-file in a different location from the current working directory, there are two options:

- Change the working directory to where the file is located.
- Add the location of the file to the *path*.

We can take the second option, and add the location to the path list, by using the `addpath` command, giving the name of the required file's location in single quotes:

```
>> addpath('/path/to/some/directory')
```

Once this is done, any script or function *m*-files that are contained in the directory can be called from the command window, no matter what the current working directory is.

■ Activity 3.7

Make a note of where you saved the function *m*-file for the code in Activity 3.6. Let's say it is called `'/path/to/file'`. *O3.E*

Now change the working directory to a different location. You can use the address bar at the top of the MATLAB environment (see Fig. 1.1), or you can type `cd /to/some/other/location` in the command window.

Try to run the instruction at the command window as before.

```
>> value = sum_function(1000)
```

You should get the following error:

```
Undefined function or variable 'sum_function'.
```

This is because the function *m*-file containing the code is not in the path.

Now add the location of the function *m*-file to the path by typing `addpath('/path/to/file')` in the command window.

Now check to confirm that the call to the function runs without an error. ■

3.8 NAMING RULES

When saving an *m*-file for a script or a function, we need to abide by the following rules on how it can be named. The name of an *m*-file

- *Must* start with a letter ...
- ... which can only be followed by
  - letters or
  - numbers or
  - underscores.
- Must have maximum of 64 characters (excluding the .m extension).
- Really should not have the same name as a MATLAB reserved word or built-in function (e.g. `for` or `sin`).

For example, names that start with a number, or contain dashes or spaces cannot be used.

For *function m*-files, the name of the *m*-file should match the name of the function it contains (the main one if it contains more than one). If they do not match, then MATLAB may give a warning (depending on the version used). For example, if the following function is saved to a file called *give_random.m* then there will be a mismatch between the function definition and the filename.

```
function result = get_random()
% Usage get_random: Return a random number
result = rand;
end
```

3.9 SCOPE OF VARIABLES

The *scope* of a variable refers to the parts of a program or the MATLAB environment where the variable is 'visible', i.e. where it can be assigned to, read, or modified. Scope is an important concept in most programming languages.

The scope of a variable that is inside a MATLAB function is described as *local*. This means that it can only be accessed and modified by commands inside the same function. It cannot be affected by, for example, instructions that are typed at the command window or instructions inside different functions or scripts.

■ Activity 3.8

Using the editor window, save the following into a function *m*-file called *my_cube.m*.

O3.A, O3.C

```
function y = my_cube(x)
% my_cube(x) : returns the cube of x

y = x * x * x;
fprintf('The cube of %u is %u\n', x, y);
end
```

Now, in the command window, assign values to variables called `x` and `y` and then call the function `my_cube`. For example, use the following commands:

```
>> x = 4;
>> y = 7;
>> my_cube(12)
>> disp([x, y])
```

What output do you expect to be produced by the call to `my_cube(12)` above?

What output do you expect from the subsequent call to `disp([x,y])` that follows? ■

The important point about the code in Activity 3.8 is that there are two variables called `x` but that they are in different scopes. The same applies to the variables called `y`. We assign values of 4 and 7 to the variables `x` and `y` in the command window. We then make a call to the function with an argument of 12. The `fprintf` command in the function will produce the following output

```
>> my_cube(12)
The cube of 12 is 1728
```

which means that the variables `x` and `y` in the scope of the *function* have values of 12 and 1728.

After the function is complete, the call to `disp([x,y])` will produce the following output for the variables in the scope of the *command window*:

```
>> disp([x y])
     4     7
```

We can see that these variables were unchanged by the call to `my_cube`.

In summary, for this example, the variables `x` and `y` that we created in the command window are not considered to be the same as the variables `x` and `y` in the `my_cube` function – because they have different *scopes*.

The scope of a variable in a function is also local when we use more than one function as shown below.

■ Activity 3.9

O3.A, O3.C

Save the following code into a function *m*-file called *log_of_cube.m*. This function contains a call to the `my_cube` function from Activity 3.8

```
function log_of_cube(y)
% log_of_cube(x) : return the log of the cube of x
z = log(my_cube(y))
fprintf('Input: %f\n', y);
fprintf('Output: %f\n', z);
end
```

Use the command window to test this function by calling it with various inputs. For each call, determine what the value is for the variable called `y` in the scope of the function `log_of_cube`. Do the same for the variable called `y` in the scope of the function `my_cube`. ■

Calling the `log_of_cube` function will create a variable `y` that is local to the `log_of_cube` function and pass this as an argument to the function `my_cube`. The function `my_cube` also has a local variable named `y` (as an output variable) but this is distinct from the one in the function `log_of_cube`. The value it has in the *calling* function does not affect the value it has in the *called* function.

Be Careful About Script *m*-Files and Scope

We have seen some of the rules for the scope of variables in functions. Note that these rules *do not apply to script m-files*.

■ Activity 3.10

O3.B, O3.C

Confirm this by saving the following command into a script *m*-file called *my_script.m*.

```
a = sqrt(50);
```

Now, return to the command window, set the value of a variable called `a`, and then call `my_script`, i.e.

```
a >> a = 4;
>> disp(a)
     4
```

```
>> my_script
>> disp(a)
     7.0711
```

Try this activity again but starting with different assigned values to a and following up with a call to my_script. ■

For the above, you should see that the variable we set in the scope of the command window *is altered by running the script*. This would not have happened if we called a *function*, even if that function also had a local variable with the name a within it.

Altering a variable in a different scope can be an unintended consequence of using scripts which generally makes using functions (rather than scripts) a safer option.

3.10 RECURSION: A FUNCTION CALLING ITSELF

We can define some functions *recursively*. This means that, while the function is running, it can call itself again, which in turn calls itself again and so on.

A good illustration of a recursive definition for a function can be obtained using the *factorial* function. The factorial of an integer is denoted with a ! sign and is equal to the product of all the numbers between 1 and the integer. The factorial of zero is defined as 1 and the factorials of the first few integers are:

$0! = 1$

$1! = 1$

$2! = 2 \times 1 = 2$

$3! = 3 \times 2 \times 1 = 6$

$4! = 4 \times 3 \times 2 \times 1 = 24$

$\vdots$

This shows that when we evaluate the factorial for a number we carry out nearly all the same operations that are needed for evaluating the factorial of the previous number. For example $4! = 4 \times 3!$. This fact is what we will use to write a recursive definition for the factorial function.

■ Activity 3.11

Use the editor window to save the following code in a function *m*-file called *my_factorial.m*. *O3.A, O3.F*

```
function result = my_factorial(n)
% my_factorial(n) : return the factorial of n
```

```
if (n<=0)
  result = 1;
else
  result = n * my_factorial(n-1);
end

end
```

This code shows that *the function calls itself* and this is what makes it a recursive function. If it is called, for example, with `n = 4`, then the result will be the value of `n` (which is 4) multiplied by the result that is obtained from calling the same function with a value of `n-1`, which is 3.

Test the function by calling it from the command window to check that it is working correctly.

Modify the function code by inserting a line to display what the value of `n` is at the start of the function's body, just before the `if` clause. What output do you obtain when calling the function from the command window?

Move the display line from the start of the function to the point just before the `end` of the function – how does the output change when it is called? ■

When the `my_factorial` function is called with an input argument greater than one, e.g.

```
>> my_factorial(3)
```

this will lead to the second part of the `if-else` clause being executed:

```
...
  result = n * my_factorial(n-1);
...
```

with a value of $n = 3$. In other words, a call is made *to the same function* passing an argument of $n - 1 = 2$. The return value from that call is multiplied by 3 to give the final result.

Each call for a value of $n > 0$ will involve a call to the same function with a value $n - 1$. Thus, the value passed as input to the function will decrease by one each time. Eventually we will have $n = 0$. In this case, and only in this case, the function will actually run the first part of the `if-else` clause:

```
...
  result = 1;
...
```

and a value of 1 will be returned and *no further recursive calls will be made*. Then the return value will be passed to the calling function, which will multiply the result by *its* local value of n and return the result to its calling function, which will multiply the result by *its* local value of n, and so on.

The first part of the `if` clause above is important – it runs if $n \leq 0$ and prevents further recursive calls. This is known as a *stopping condition* for the recursive function. Stopping conditions are needed to prevent a recursive function from calling itself infinitely. A stopping condition is also known as a *ground case*.

3.11 SUMMARY

Functions can be defined by the user in MATLAB, and they can be very powerful. They allow programmers to expand the basic capabilities of MATLAB for their own specific purposes. They can be saved in files in the same way as scripts but one needs to be careful regarding the differences between scripts and functions.

Functions should be saved in a file with the same name as the function (with an added *.m* suffix). If the file is saved in a different location from the current working directory, then that location must be added to the *path* that MATLAB uses to search for *m*-files.

A function *m*-file can also contain *sub-functions* or *helper* functions that are called by the 'main' function in the file.

Functions take input values (arguments) and return values. Variables defined inside functions have a *scope* that is *local* to the function and this means that they cannot be accessed from outside the function, for example from the command window or from other functions.

MATLAB allows definition of *recursive* functions. Recursive functions are functions that make calls to themselves. Recursive functions need a stopping condition to prevent them from calling themselves infinitely.

The `error` command can be used to report errors within functions, for example by checking that the input arguments have the correct size or data type.

3.12 FURTHER RESOURCES

Look at the *File Exchange* section of the Mathworks web site (the company that develops MATLAB). You will find a large number of useful MATLAB functions that have been written and shared by other people. This can be an extremely valuable time-saving resource. The URL is: http://www.mathworks.com/matlabcentral/fileexchange/.

EXERCISES

■ Exercise 3.1

O3.A

Identify problems and errors in the following functions that relate to their input and output arguments.

1. func1:

```
function f = func1(x, y)
x = 3 * y;
end
```

2. func2:

```
function f = func2(a, b)
f = sqrt(3 * a);
end
```

3. foo:

```
function [a, b] = foo(x, y)
b = a + x + y;
end
```

4. findSquareRoot:

```
function findSquareRoot(x)
% Usage : findSquareRoot(x)
%   Provide the square root of the input
returnValue = sqrt(x)
end
```

■

■ Exercise 3.2

O3.A

In Exercise 2.12 you wrote a script *m*-file to compute the Injury Severity Score (ISS) given an array of 6 Abbreviated Injury Scale (AIS) values. Modify your solution so that it uses a *function m*-file rather than a *script m*-file. Specifically, the function should take an array of 6 AIS values as its only input argument, and produce the ISS value as its output. As described in Exercise 2.12, the ISS is computed from the AIS values as follows:

- Find the three highest AIS values in the array.
- Square their values.
- Sum the squares.

Also, write a short *script m*-file that defines the AIS array `[3 0 4 5 3 0]` and contains a call to your function using this array as an argument. The correct ISS value for the array should be 50. ■

■ Exercise 3.3

O3.E

Modify your solution to Exercise 3.2 so that the function *m*-file is contained in a different location in the file system. For example, make a sub-directory of the directory containing your script *m*-file, and move the function *m*-file into it. Then, modify your script so that it is still able to find the function *m*-file. ■

■ Exercise 3.4

O3.A, O3.C

Examine the following code which is saved in the file *fn1.m*.

```
function x = fn1(y)

x = fn2(y + 2);
disp([x,y]);

end

function x = fn2(y)

x = y^2;
disp([x,y]);

end
```

The file is available to you through the book's web site.

Without using MATLAB, predict what the output of the function would be if the following call was typed in the command window:

```
>> fn1(10);
```

Verify your prediction by downloading the function *m*-file and making the call. ■

■ Exercise 3.5

O3.A, O3.D

A function takes three numeric arguments representing the coefficients of a quadratic equation and returns two values which are the two roots of the equation. The name of the function is `quadRoots`.

1. Write a suitable prototype or definition for the function. Add an appropriate help and usage section at the top of the function and save it in a suitably named file.
2. Add a test to check if the quadratic roots are complex and to return with an error if this is true (see Section 3.3, `doc error` or `help ... error`).
3. Complete the function `quadRoots` and test that it gives the correct results.
4. Modify the `quadRoots` function so that it accepts a single array containing all three coefficients and returns a single array containing the roots. ■

■ Exercise 3.6

O3.A, O3.B, O3.D

In Exercise 2.8 you wrote a script *m*-file to automatically identify troughs in gait tracking data. As well as identifying troughs in such data it can be useful to identify peaks too. Recall that a trough can be defined as a value that is less than the value(s) before and after it. Similarly, a peak can be defined as a value that is greater than the value(s) before and after it.

First, convert your code from Exercise 2.8 to use a function *m*-file. The function should take a single argument, representing the array of values to be searched, and return two values: an array of indices of peaks and an array of indices of troughs. The main script *m*-file should load in the same data file that you used in Exercise 2.8 (*LAnkle_tracking.mat*), display the plot of *z*-coordinates over time as in Exercise 2.8, and then call the new function to determine the peaks and troughs. Finally, the script should use the indices returned to determine the timings of all peaks and troughs. ■

■ Exercise 3.7

O3.A, O3.B, O3.D

X-ray computed tomography (CT) scanners produce images whose intensities are in *Hounsfield Units* (HU), named after the inventor of the CT scanner, Sir Godfrey Hounsfield. The HU scale is an approximately linear transformation of attenuation coefficient (AC) measurements into one in which the radiodensity of distilled water at standard pressure and temperature is defined as zero HU, while the radiodensity of air is defined as −1000 HU. When reconstructing Positron Emission Tomography (PET) images the reconstructions need to be corrected for the effects of attenuation using the attenuation coefficients. Therefore, there is a need to convert from HU to AC. The figure below illustrates such a relationship, reported in [3].

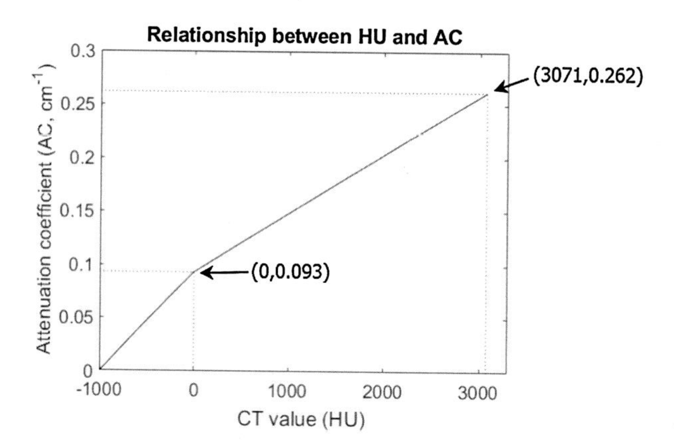

Using the relationship illustrated in the figure, write a MATLAB function to convert HU to AC. Your function should check if the HU value is between

−1000 and 3071 and generate an error if it isn't. Write a script *m*-file to read an HU value from the user, convert it into AC and display the result. ■

■ Exercise 3.8

O3.A, O3.B, O3.D

A company is trialling a portable personal device that monitors a range of clinical indicators. One of these is heart rate, and the company would like to embed some software in the device to raise an alert if a sudden change occurs in the heart rate.

The file *heart.txt* contains sample heart rate data, acquired once per minute from a patient who was wearing the new device. Your task is to write a MATLAB script *m*-file that reads this data into an array, and searches for any sudden changes in heart rate.

A sudden change is defined as a heart rate that is more than 20% different (i.e. larger or smaller) than the average of the previous five heart rate values. Therefore, you should only start to look for sudden changes from the 6th measurement onward. When your program identifies a sudden change it should display a warning such as:

```
Warning - abnormal increase in heart rate at 123 minutes
```

As the heart rate measurements are made once per minute you can simply use the array index of the abnormal value as the time of the sudden change. Note that the program should indicate whether the heart rate change was an increase or a decrease. All sudden changes within the array should be reported. If no sudden changes are detected the following message should be displayed:

```
No abnormal changes detected
```

As part of your solution you should define a function *m*-file that takes an array of heart rate values as its only input argument and returns as its output an array of the indices of all sudden changes and also values indicating whether each change was an increase or decrease. If the function is provided with an empty array it should exit with an appropriate error message.

A separate script should call your function and should also take responsibility for reading the data (see Section 1.6) and printing the output message(s). ■

■ Exercise 3.9

O3.A, O3.B

Many medical imaging modalities acquire imaging data in a number of acquisitions over time. *Gating* is a common technique that aims to reduce artefacts in the resulting images that are due to cyclic motion (such as heart beats). The approach involves accepting only data that were acquired at approximately the same motion position and discarding the rest. One way to decide whether to accept data or not is to use a simple signal that is related to the motion of interest (for example an electrocardiographic signal, or ECG). The percentage of data accepted, out of all those acquired, is known as the *scan efficiency*.

A *bellows* can be used to generate a signal for *respiratory gating* of magnetic resonance (MR) data. It is an air-filled bag that is strapped around the chest. As the subject breathes, the air moving into and out of the bag is measured by a sensor. The file *bellows.mat* contains a set of respiratory bellows data.

For respiratory gating using the bellows signal, we need to determine a pair of signal values: a low value and a high value. Only data with a signal between these values will be accepted. The low and high values define the *gating window*.

The bounds of the gating window are typically determined using the signals acquired during a short 'preparation phase'. One way to set the high value is simply to use the maximum value observed during the preparation phase. The low value then needs to be set so that the percentage of data accepted during the preparation phase is equal to the desired scan efficiency.

Write a MATLAB function *m*-file to determine the bounds of a gating window given an array of bellows data acquired during a preparation phase and a desired scan efficiency. Write a short script *m*-file to read in the data in *bellows.mat* and call the function treating the first 5000 measurements as the signals from the preparation phase. Use a desired scan efficiency of 25%.

Once the gating window has been determined, your script should compute the indices of all elements in the full bellows array that are within the gating window. The script should then compute and display the *final* scan efficiency, i.e. the percentage of the full set of bellows signal values that lie within the gating window. ■

■ Exercise 3.10

O3.A, O3.F

The Fibonacci sequence is 1, 1, 2, 3, 5, 8, ... It is defined by the starting values $f_1 = 1$, $f_2 = 1$ and after that by the relation $f_n = f_{n-1} + f_{n-2}$. The Fibonacci sequence has many applications in a wide range of fields including in biology [4].

Write a *recursive* function to find the nth term of the Fibonacci sequence. Use the following function definition and ensure that the code has a help section and good comments.

```
function f = fibonacci(n)
```

■

■ Exercise 3.11

O3.A, O3.F

A good example of working recursively is given by the evaluation of a polynomial.

A polynomial is defined by its coefficients. As an illustration, assume we have a cubic polynomial with the coefficients 2, 3, 7, 1, i.e. the polynomial $2x^3 + 3x^2 + 7x + 1$. If we evaluate this expression directly it requires 3, 2 and 1 multiplications for the first three terms and three additions to give a total of nine operations. We can, however, write the evaluation in a recursive way that requires fewer operations: If we take out successively more factors of x from the terms that contain them we get the following:

$$\begin{aligned} 2x^3 + 3x^2 + 7x + 1 &= x(2x^2 + 3x + 7) + 1 \\ &= x(x(2x + 3) + 7) + 1 \\ &= x(x(x(2) + 3) + 7) + 1 \end{aligned}$$

The last line shows how the polynomial may be evaluated with three additions and three multiplications giving a total of six operations, three fewer than the direct approach. This illustrates one of the benefits of working recursively.

The above suggests how a function to evaluate a polynomial may be written recursively. An evaluation of the above polynomial could therefore be written as a sequence of calls to a function f that starts off by taking an array containing the coefficients.

$$2x^3 + 3x^2 + 7x + 1 = \quad f([2, 3, 7, 1], x)$$

$$\begin{aligned} f([2, 3, 7, 1], x) &= x\, f([2, 3, 7], x) + 1 \\ f([2, 3, 7], x) &= x\, f([2, 3], x) + 7 \\ f([2, 3], x) &= x\, f([2], x) + 3 \\ f([2], x) &= 2 \end{aligned}$$

Write a *recursive* MATLAB function to evaluate a polynomial with a given set of coefficients and a given x value. The definition of your function should be

```
function y = my_poly_value(coeffs, x)
```

Make sure that the code is well commented and that there is a help section at the top. Also make sure that you include a stopping condition for when the array of coefficients contains a single element.

Use your function to find the value of y when $x = -2$ for the polynomial $x^4 - 2x^3 - 2x^2 - x + 3$. You might also want to check that your function agrees with the built-in function `polyval`. Type `doc polyval` for details of how to use this function. ■

FAMOUS COMPUTER PROGRAMMER: BETTY JEAN BARTIK

Betty Jean Bartik (née Jennings) was an American mathematician who worked at the University of Pennsylvania in the 1940s. She was one of a team of six women who were responsible for programming the ENIAC, which was the world's first electronic digital computer. Along with her colleagues Kay McNulty, Betty Snyder, Marlyn Wescoff, Fran Bilas and Ruth Lichterman, Betty Jean Bartik programmed the ENIAC by physically modifying the machine, i.e. moving switches and rerouting cables, rather than using a symbolic language such as those that we are used to today.

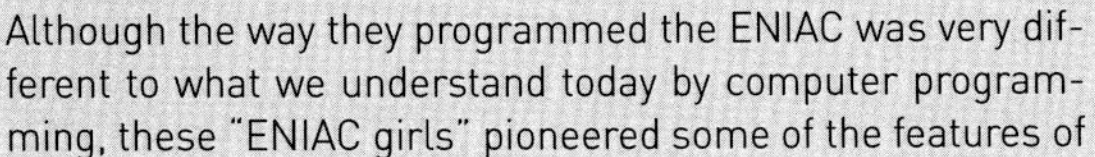

Although the way they programmed the ENIAC was very different to what we understand today by computer programming, these "ENIAC girls" pioneered some of the features of the programming languages we use now, such as subroutines (i.e. functions) and nesting. Arguably they invented the discipline of programming modern digital computers.

However, for a long time these women did not receive the recognition they deserved. Greater credit was given to the team that built the computer's hardware, rather than the team that programmed it. It was only many years later that historians of computing began to appreciate the contributions of these women. In 2008, two years before her death, Betty Jean Bartik received a Pioneer Award from the IEEE Computer Society and became a fellow of the Computer History Museum in California, USA.

> *"I was told I'd never make it to VP rank because I was too outspoken. Maybe so, but I think men will always find an excuse for keeping women in their 'place.' So, let's make that place the executive suite and start more of our own companies."*
>
> **Betty Jean Bartik**

CHAPTER 4

Program Development and Testing

LEARNING OBJECTIVES

At the end of this chapter you should be able to:

O4.A Incrementally develop a MATLAB function or program
O4.B Test each stage of the incrementally developing function or program
O4.C Identify common types of error that can occur in MATLAB functions and scripts
O4.D Debug and fix errors in scripts and functions
O4.E Use MATLAB's built-in error handling functions

4.1 INTRODUCTION

Most tasks in realistic computer programming settings involve writing substantial amounts of code. Such programs typically consist of functions or libraries that can call each other in complex ways. Furthermore, tackling large computer programming tasks can often involve more than one person collaborating and working on the same code. Therefore, there is a need to work methodically and break a large project down into manageable and testable portions.

We already saw some of the debugging and code analysis tools that MATLAB has to offer (Sections 1.12 and 1.12.2) and other languages such as C++, Python or Java also have similar debugging and analysis tools. In this chapter we will revisit these tools and discuss a general approach for developing and implementing a significant programming project.

4.2 INCREMENTAL DEVELOPMENT

Working on the principle of breaking a large task down into smaller, more manageable and (importantly) testable sub-tasks, the process of *incremental development* can be useful when trying to solve a difficult task or model a complex system using a computer program.

MATLAB Programming for Biomedical Engineers and Scientists. DOI: 10.1016/B978-0-12-812203-7.00004-5

When programming something substantial, it is better to avoid trying to take a "big bang" approach to coding. We will generally fail if we try to solve it all at once.

General tips for incremental development are:

- Write code in small pieces.
- The pieces may well be incomplete but should work without error.
- Don't try and write everything in one massive function in one go.
- For example, if the program needs to perform some user interaction, a typical incremental development process might involve the following:
  - Start off with just enough code to display output.
  - Run that part and check it.
  - Then write a bit more code to ask the user for input.
  - Check the new part by printing to the screen what the user provides as input.
  - Add a bit more code to actually do something (simple) with the user's input.
  - Check the 'doing something simple part' – perhaps by again printing out its result.
  - Add a little more code to do something a bit more than the simple stuff already done.
  - etc.
  - Continue these steps of adding a bit more code and checking it works until we have a final working version of the program.

This kind of iterative way of working with frequent checking is important. Adding a large amount of code before checking can be difficult. First, the program itself may not run. Second, it may give incorrect output. In either case, if a large amount of code was introduced since the last check it will be more difficult to find where the bug or error is.

The bottom line is that we should aim to always have a version of our program that runs without error.

This chapter tries to illustrate this principle by example. We first describe the task in the next example and after that, one way to take the incremental steps of building and testing some code to carry out the task is given.

■ Example 4.1

O4.A, O4.B, O4.C, O4.D

This example will look at data representing measurements of heart contraction (see Example 1.2). In a healthy heart, all parts of the cardiac muscle (the *myocardium*) will contract in time with each other. *Cardiac dyssynchrony* occurs when there are differences between the timings of the contractions of different parts of the myocardium. Such differences may indicate the presence of heart disease.

One form of dyssynchrony is *intraventricular dyssynchrony*, which occurs when the timing differences are between the contractions of different segments of the myocardium of the left ventricle of the heart. A common way of defining the different parts of the left ventricle, in order to assess for dyssynchrony, is to use the American Heart Association's (AHA) 17-segment model.

Using echocardiography or an MR scanner, it is possible to measure the radial displacement of each segment over time and these measurements form the starting data for the analysis. There are a number of ways of measuring intraventricular dyssynchrony and one method is to use the standard deviation of the *time-to-peak* radial displacement of the AHA segments. Using the radial measurements, the time between the start of the cardiac cycle and the peak radial displacement is determined for each segment. The standard deviation of these times is the indicator of dyssynchrony.

The task is to write a MATLAB function to compute the standard deviation of time-to-peak displacements. The input to the function should be an array containing radial displacement data broken down by AHA segment. The array for a patient contains 17 columns, one for each segment. Each column contains the displacement measurements for the corresponding segment. Sample input arrays for three patients are provided in the file *radial_displacements.mat*, which can be used to test the function. Loading this file will create arrays with names `patient1`, `patient2` and `patient3`. The *temporal resolution* (i.e. the time between displacement measurements) of each segment's data is 30 milliseconds.

The function should first determine the index of the maximum radial displacement for each segment. As the temporal resolution is known, this index can be converted to give the time-to-peak value for the segment. When the times to peak are calculated for all segments, their standard deviation can be computed and returned as the output of the function.

A script should load the test data and, for each patient, it should call the function, display the standard deviation value, and display whether or not the patient has 'significant' dyssynchrony. We consider a patient to have significant dyssynchrony if their time-to-peak standard deviation is greater than 60 milliseconds. ■

Step 1

In keeping with the principles of incremental programming, we start by writing a very small and basic script:

```
clear; close all;

% load data
load('radial_displacements.mat');
```

```
% display results
disp('Inside the main script!');
```

Note how this script does very little of what we ultimately aim to do. We can save this in the file *main_script.m*.

Test 1

It does not matter that the script does very little, we should still run this script from the command window to test it:

```
>> main_script
Inside the main script!
```

All our script does so far is to clear the workspace, close any open figures, load in the measurement data, and give a dummy message. Once we are happy that it does this, we can move on to the next step.

Step 2

Now we start writing the function that will do the work. We can use the editor to create a file *std_dev_ttp.m*. We begin with the following:

```
function ttp_SD = std_dev_ttp(rad)

% Standard deviation of time-to-peak, return value.
ttp_SD = 100;

end
```

It is okay that our function does not yet do what we want – the function still has some important features: It has the correct *prototype*, i.e. it expects an argument of radial displacements as input and says that it will return an output value indicating the required standard deviation (even though this is currently hard-coded to a 'dummy' value of 100).

Now that we have written a (very limited) function, we can adjust the main script to call it. We can pass one of the arrays loaded from the data file to the function and collect the result that is returned:

```
clear; close all;

% Load data
load('radial_displacements.mat');

% Call the function.
sttp1 = std_dev_ttp(patient1);

% Display results
disp([' Std dev of time-to-peak displacement = ' num2str(sttp1) 'ms']);
```

Test 2

Now we can check the output of the code by once again calling the main script at the command window:

```
>> main_script
  Std dev of time-to-peak displacement = 100ms
```

The test gives the output we expect: the function is called by the main script and it is currently hard-coded to return a value of 100 ms. This returned value is then displayed as was required.

Step 3

Now we can add a little more functionality to our main script. We could include a check to see if the dyssynchrony measure is safe or not and display an appropriate message. We do this by modifying *main_ script.m* as follows:

```
clear; close all;

% load data
load('radial_displacements.mat');

% Call the function.
sttp1 = std_dev_ttp(patient1);

% display results
disp(['  Std dev of time-to-peak displacement = ' num2str(sttp1) ...
     'ms']);

if (sttp1 < 60)
    disp('  Patient is dyssynchronous');
else
    disp('  Patient is not dyssynchronous');
end
```

Here we have added a few lines of code at the end (containing an `if-else` clause) to check if the value is above 60 or not.

Test 3

We test the code as before:

```
>> main_script
  Std dev of time-to-peak displacement = 100ms
  Patient is not dyssynchronous
```

Here we can see that the code does *not* work as expected – we have found a bug! The function currently is hard-coded to return a value 100 ms, which is above the limit of 60 ms for determining dyssynchrony, but the message displayed says that the patient is not dyssynchronous. In order to fix this, we only need to look back over the (small) amount of code we just added. On reviewing the `if-else` it is clear that the test `if (sttp1<60)` is the cause of the problem and the bug can be fixed easily by re-writing it as `if (sttp1>60)` instead.

A second quick and easy test at this stage is to modify the hard-coded value that the function returns, for example to a value of 30 ms. We modify *std_dev_ttp.m* as follows,

```
function ttp_SD = std_dev_ttp(rad)

% Standard deviation of time-to-peak, return value.
ttp_SD = 30;

end
```

and then re-run the code:

```
>> main_script
  Std dev of time-to-peak displacement = 30ms
  Patient is not dyssynchronous
```

This quick test also confirms what we expect (having fixed the bug in the main script as described above).

Step 4

In the steps so far, we have only really built a skeleton of the program we want. Now we can begin to think about the 'core' functionality that is required. We can modify the function *m*-file so that we find the peak radial displacement for each segment. To do this we will need to loop over the array of radial displacement measurements `rad`. Recall that there are 17 segments in the AHA definition and the array has a column of measurements for each segment. We need to find the index of the largest value (peak) in each column. We add code as follows:

```
function ttp_SD = std_dev_ttp(rad)

nSegments = 17;

% Loop over the 17 segments and compute time-to-peak
% displacements for each one.
for seg = 1:nSegments
   [peak, peak_index] = max(rad(:,seg));

end

% Standard deviation of time-to-peak, return value.
ttp_SD = 30;

end
```

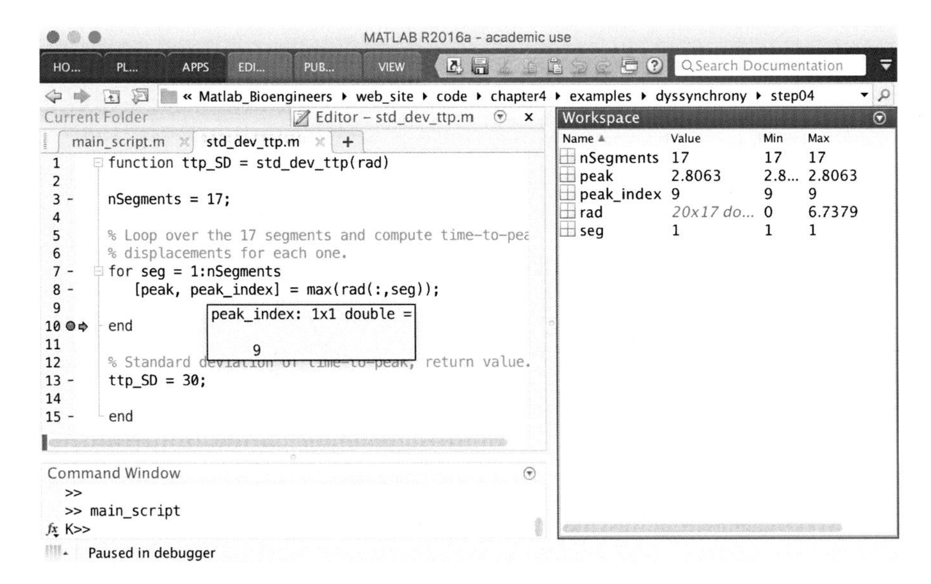

FIGURE 4.1 Using a breakpoint to inspect program variables.

Here, we have added a loop to find the index of the maximum value for each segment (column). We are not actually using this information yet, but it is a small incremental addition of code that will help us later.

Test 4

One way to check that the new code is working is to place a breakpoint in the loop that was added. Then we can call the main script and inspect variables whilst the code is running, as shown in Fig. 4.1.

Here, we have added a breakpoint at the end of the loop and the code has stopped in the first iteration (see the value of the loop variable `seg` in the workspace window on the right). We can check variables in the code by hovering the mouse over them. In the above, we can see that the variable `peak_index` has a value of 9. How can we test if this index for the peak is sensible? One way is to look at the contents of the current column in the radial measurements array, for example by simply typing it at the command window:

```
K>> rad(:,1)'
ans =
  Columns 1 through 6
         0    0.1031    0.7285    1.3087    1.7075    2.2302
  Columns 7 through 12
    2.5135    2.7005    2.8063    2.7488    2.4591    2.1285
% .... etc.
```

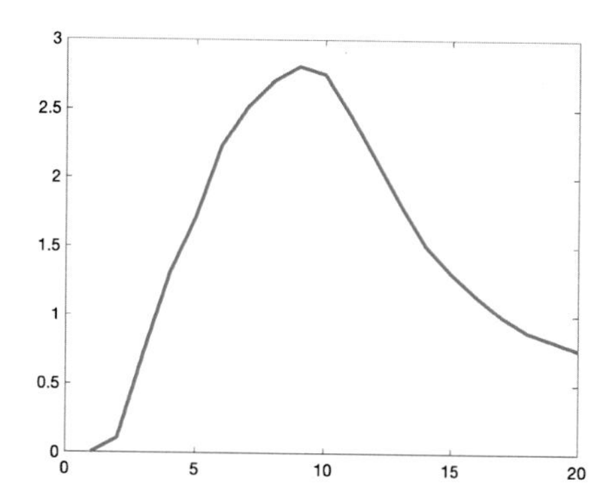

FIGURE 4.2 A plot of radial displacements for a segment of the left ventricular myocardium.

Here we can see that, at index 9, the value is 2.8063 which is indeed the highest. If the array we want to inspect is very large, displaying it in the command window is inconvenient and an alternative would be to use the `plot` command to visualize the data, i.e.

```
>> plot(rad(:,1))
```

The resulting plot is shown in Fig. 4.2 and confirms that the index we have found (9) corresponds to a peak.

The Remaining Steps

In order to complete the task, we have a few more items to add to our function's code:

- Usage and help text for the function. This can be done at any stage and could have been done right at the start.
- Code to calculate the *time* of a peak from the index that we have found so far.
- A variable to store the times for *all* the segments.
- A line or two of code to find the standard deviation of these times.

All of these items can be added incrementally in the same way as we have been working above. After each step, a quick test can be carried out to check (a) if it runs, and (b) gives sensible results/values.

■ Activity 4.1

O4.A, O4.B

Complete the task above so that the code includes the items remaining. Use the same incremental add-code-then-test approach described above. ■

4.3 ARE WE FINISHED? VALIDATING USER INPUT

Once the code for the dyssynchrony task is finished, we could decide that the task is complete as we have a working function that achieves its initial aim. However, in another sense we may still have work to do because the function might need to be made more *robust*. One way of making code robust is to protect it against possible misuse.

Validating user input is an important part of programming; one of the main reasons for program failure is when a user (or another part of the program) gives input to a function or program that does not match what the programmer expected.

In the case of Example 4.1, we could still check to see if a sensible array has been provided as input, for example it should not be an empty array perhaps. The array should also be consistent with the AHA heart segment model – in other words it really should have exactly seventeen columns.

For example, if we want to ensure that the number of columns is correct, then we can introduce the following code near the top of the function:

```
function ttp_SD = std_dev_ttp(rad)
% ... Usage and help text here

nSegments = 17;

if (size(rad,2) ~= nSegments)
  error('std_dev_ttp: Expecting exactly 17 columns in the input ...
       array.');
end

% ... rest of function as before.
```

If the function receives an array that does not have seventeen columns, it now prints out an error message and returns without continuing to the rest of the code.

Notice how the error message begins by giving the name of the function where the error is raised. This is good practice as programs can use multiple functions and it makes identifying the locations of errors easier.

If we want to give a warning message to the user but still want to carry on executing the function (rather than returning), then we can use the built-in function `warning`. Type `help warning` and `help error` at the command window for more information.

4.4 DEBUGGING A FUNCTION

The basics of the MATLAB debugger were described in Section 1.12. Here, we consider some of the types of errors that the debugger can be used to help identify and fix.

When writing a program or function that is non-trivial, the chances of it working perfectly first time are quite low, even for a skilled MATLAB user. Debugging tools are there to help identify why a program does not work correctly or as expected.

When debugging, we need to work systematically and run our program with various sets of test data to find errors and correct them when we find them. If we work incrementally as we did in Example 4.1 above, then the repeated nature of the testing allows errors to be caught earlier, which reduces the chance of later errors occurring or minimizes their severity.

There are three main types of error that a function or program can contain:

- Syntax errors.
- Run-time errors.
- Logic errors.

The term 'syntax' relates to the *grammar* of the programming language. Each language has a specific set of rules for how commands can be written in that language and these rules are collectively known as the language's syntax.

When a line in a MATLAB function contains a syntax error, the built-in *Code Analyzer* (see Section 1.12.2) should highlight in red the corresponding line of code. Hovering the mouse over the line should result in a message being displayed to the user that describes what MATLAB has decided is the particular syntax error in the line. In other words, MATLAB can detect an error with the script or function before it is run.

■ Example 4.2 (A syntax error)

O4.C, O4.D

If the following code is entered into a script, say *my_ script.m*, using the editor, then the line containing the `if` statement should be highlighted in red to indicate a problem with the syntax.

```
clear, close all

randVal = rand;

if randVal > 0.5
  disp('Greater than 0.5')
```

This is because the syntax rules of MATLAB require a closing `end` statement for each `if` statement.

If we fail to notice this error and attempt to run the script from the command window, we will get an error when MATLAB reaches the relevant part of the code and a more dramatic message will be displayed:

```
>> my_script

Error: File: my_script.m Line: 5 Column: 1
At least one END is missing: the statement may begin here.
```

Clicking on the part of the error report with the line number will take us directly to the corresponding place in the script file so that we can figure out how to correct the error. In this case, adding an `end` statement after the `disp` call should be enough. ■

■ Activity 4.2

Type the code from the previous example into a script. Confirm that the Code Analyzer highlights the error. Fix the error and confirm that the highlighting disappears. ■

O4.C, O4.D

■ Example 4.3 (Another syntax error)

The following code contains a syntax error. The MATLAB code analyzer should highlight it in red. The code begins by generating a single random number between 1 and 10 and then it seeks to decide if the random number equals three or not, adjusting its output in each case.

O4.C

```
clear, close all

a = randi(10, 1);

if a = 3
  disp('Found a three!')
else
  disp('Not a three')
end
```

■

■ Activity 4.3

Type the code from the previous example into a script. Confirm that the Code Analyzer highlights the error. Fix the error and confirm that the highlighting disappears. ■

O4.C, O4.D

The reason for this syntax error is that the = sign used in the `if` statement is an *assignment operator* and should not be used to test for *equality*. The correct symbol to use here is the *comparison operator* which is == in MATLAB, i.e. two equals signs, not one (see Example 2.2). This is also the case in quite a few other languages and it is very common to mix up these operators.

■ Example 4.4 (A run-time error)

O4.C

Some errors cannot be determined by MATLAB while we are editing the code and can only be found when the code is run. This is because a specific sequence of commands needs to be run to bring about the conditions for the error. Consider the following code in a script called *someScript.m*. It aims to generate three sequences of numbers and assign them to array variables a, b and c. Then it tries to find the sums of each of these sequences (line numbers have been shown here to help the description).

```
1   clear, close all
2
3   a = 1:10;
4   b = 2:2:20;
5   c = 3:3:30;
6
7   sumA = sum(a);
8   disp(sumA)
9
10  sum = sum(b);
11  disp(sum)
12
13  sumC = sum(c);
14  disp(sumC);
```

In this case, there are no syntax errors but running the script at the command window leads to a run-time error as follows:

```
>> someScript
    55

   110

Index exceeds matrix dimensions.

Error in someScript.m (line 13)
sumC = sum(c);
```

■

■ Activity 4.4

O4.C

Type the code from the previous example into a script. Run the script from the command window to reproduce the error. ■

The problem with the code in Example 4.4 is related to the choice of variable names. The sums of the arrays a and c have been given sensible names (sumA and sumC). The sum of the array variable b has been stored in a variable called sum. This is a problem because the local script has used the name of a *built-in MATLAB function* as a name for a local variable.

Using the name of a built-in function for a variable that we use locally is called *shadowing*. We can check if a name is already used by MATLAB or in the workspace using the `exist` command. In this case, we can trivially check

```
>> exist('sum')

ans =
     5
```

The return value of 5 indicates that the name `sum` exists and is attached to a function (as opposed, say, to a file). Type `help exist` at the command window for more details. We can check a more sensible name as follows:

```
>> exist('sumB')

ans =
     0
```

The value of 0 returned indicates that the name `sumB` is not being used and we can safely use it in our code.

Returning to the code in Example 4.4, when the execution reaches line 13, the programmer's intention was to use the built-in MATLAB function called `sum`, but this now refers to a local variable that contains a single number. The use of the brackets in line 13 is now interpreted as a request for an element or elements taken from an array called `sum`. The elements requested are at the indices specified by another variable, `c`. That is, the indices requested start at 3 and go up to 30 in steps of 3. The problem however, is the variable called `sum` is just a single scalar value, i.e. it as an array of length one. It is not possible to obtain an element from it at index 3 or higher. This is the reason for the code crashing at run-time.

Apart from giving an example of a run-time error this example shows why it is very important *not* to give names to variables that are already used to refer to built-in functions. MATLAB will not complain when you do this so be careful!

■ Example 4.5 (A run-time error)

Here is some more code that gives a run-time error:

O4.C

```
clear, close all

a = 60 * pi / 180;

b = sin(a + x)
```

In this case, when the code is run, the following output is given

```
>> someScript

Undefined function or variable 'x'.

Error in someScript.m (line 5)
b = sin(a + x);
```

■

■ Activity 4.5

O4.C

Type the code from Example 4.5 into a script. Run the script from the command window to reproduce the error. Can you see why this error occurs? ■

The script in Example 4.5 calls `clear` at the start. This means that even if the workspace contained a global variable called `x` before calling the script, then it would have been cleared and would no longer be available. Hence, the request to calculate `sin(a + x)` will always fail as there is no variable called `x`. This would need to be defined before the point where the `sin` function is called to make the code work (but after the `clear`).

■ Activity 4.6

O4.D

Fix the code from Example 4.5 by introducing an appropriate local variable and assigning a value to it. Run the script to check it works. ■

Logic Errors

Logic errors can be some of the hardest to find, and identifying them often requires using the debugger to carry out careful line-by-line inspection of the code as it is running. This is because no error is reported by MATLAB and the Code Analyzer does not highlight anything either. However, when the code is run, an incorrect output or behavior results and MATLAB will generally not provide a clue as to why this happened.

■ Example 4.6 (A logic error)

O4.C

In this example, the following code attempts to show that $\sin\theta = \frac{1}{2}$ when $\theta = 30°$:

```
clear, close all

theta = 30;
value1 = sin(theta);
value2 = 1/2;

fprintf('These should be equal: ')
fprintf(' %0.2f and %0.2f\n', value1, value2)
```

When we actually run the code, we get the following output with no error reported:

```
>> someScript
These should be equal: -0.99 and 0.50
```

■

This shows two numbers that are clearly not equal so what has gone wrong?

■ Activity 4.7

Type the code from Example 4.6 into a script and run it to reproduce the same output.

O4.C, O4.D

Can you see why the output is incorrect? Place a breakpoint in the code and run the script again. Check the values of the variables to see if there is a clue. ■

To track down the reason for the behavior in Example 4.6 we need to carry out some further investigation. In the command window we can look at the help on the `sin` function which we might suspect of being broken:

```
>> help sin
 sin    Sine of argument in radians.
    sin(X) is the sine of the elements of X.
```

This shows that the built-in MATLAB trigonometric function `sin` expects its argument in *radians* so that is what we should give it. The function when used above gave the *correct result* for the sine of 30 radians – not degrees. That is, there was a mismatch between our expectation of the function and what it was designed to do. This error is readily fixed in a number of ways, for example by replacing `theta = 30;` with `theta = 30 * pi / 180;` where we have made use of the built-in MATLAB constant `pi`.

■ Activity 4.8

Fix the code in Example 4.6 and confirm that $\sin 30^\circ$ does indeed equal 1/2! ■

O4.C, O4.D

Because no error message is given, logic errors can be difficult to detect. Sometimes we might not even notice that the program or function is behaving wrongly. Even when we suspect a logic error exists, by noticing something incorrect in the program's behavior, it can still be difficult to pinpoint where the error is. There is no error message helpfully pointing us to a specific line in the code, as in the case of syntax errors. Worse still, the error will not necessarily be restricted to just one line – it can arise from the way different sections of code interact.

The error in the previous example was due to the intention to use degrees where radians should have been used instead. This is only noticed by observing the actual values of the output and seeing that the numeric value is not correct; in other words, by careful inspection of the output.

One way to identify logic errors in a function is to use the MATLAB interactive debugger to *step through the function line-by-line*, perhaps with a calculator or pencil and paper to hand (see Section 1.12).

Another way is to take each line of the code and copy-and-paste it into the command window, executing each line at a time and assessing its result to see if it is right.

Putting lots of statements in the code that display output and the values of the most relevant variables using the `disp` or `fprintf` commands can also help. Values of variables can be displayed before and after critical sections of the code, e.g. calls to other functions or important control sections such as `if` clauses and `for` or `while` loops. Once we have checked and are satisfied that the function or program is working correctly, the extra output statements should be removed.

4.5 COMMON REASONS FOR ERRORS WHEN RUNNING A SCRIPT OR A FUNCTION

There are a few common reasons for errors when running a script or a function. These include:

- We use matrix multiplication when we meant to use *element-wise* multiplication. For example, the following code

  ```
  x = [1 2 3];
  y = [6 7 8];
  z = x .* y;
  ```

  multiplies `x` and `y` *element-wise* to give a result of [6, 14, 24]. It is a common mistake to write `z = x * y` instead of the above which would lead to an error. In this case it would lead to a run-time error because it is not possible to matrix-multiply the values of `x` and `y` as defined above (because the sizes don't match).
- The location of the code that we want to run is not in the MATLAB path, or some other code that it relies on is not in the path (see Section 3.7).
- We forget that MATLAB is *case-sensitive*, e.g. the variables `x` and `X` are different, as are `myVar`, `myvar` and `MyVar`.
- We accidentally set off an infinite loop, for example:

  ```
  count = 0; sum = 0;
  while(count < 100)
    sum = sum + count;
  end
  ```

will start but never end because the variable `count` never gets updated within the loop. The following is one way to fix this:

```
count = 0; sum = 0;
while(count < 100)
  sum = sum + count;
  count = count + 1;
end
```

- We have exceeded the memory available for our computer. For example, the command `rand(N)` returns a matrix of random numbers in an array of size $N \times N$. So if we run the following code

```
N = 100000;
R = rand(N);
```

then the computer could get very 'tired' as it runs out of memory trying to create a $10^6 \times 10^6$ array. This corresponds to around 3 or 4 terabytes if four bytes are needed for each array element. This can be viewed as a run-time error.

4.6 ERROR HANDLING

In the preceding sections, we discussed some of the kinds of error that may occur when writing code. Some of these can be hard to predict but it is still a good idea to write code in a way that anticipates the kinds of errors that might occur and tries to deal with them in advance to reduce any disastrous consequences. This can be done through *error handling* and we give some illustrations in this section of how this can be done.

In general if there is a run-time error that is encountered when the program is run, then MATLAB itself will stop running the program and return. It will give an error message (see Example 4.5), but sometimes the error message that MATLAB gives can be uninformative and we might want to make the error reporting more useful and specific.

4.6.1 The `error` and `warning` Functions

We first introduced the `error` function in Section 3.3, and in Section 4.3 we saw a way of using it whilst validating user input. This function, and the closely related `warning` function, can be useful when testing for correct behavior when code is running. Once an error is detected, we can use them to give tailored messages to the user.

■ Example 4.7

For example, we may have some function that is expected to carry out some processing and save a record of its activity to a given file. For this to work the file may be required to already exist:

O4.E

```
function processFile(logFileName)

if ( ~ exist(logFileName, 'file') )
  msg = sprintf('No file: %s. Bailing out!\n', logFileName);
  error('MATLAB:processFile:fileMissing', msg)
end

% further code below
```

This code first tests to see if the log file with the given name exists. If it does not, a specific error message saying this is generated using the `sprintf` function. The message is contained in a string variable called `msg` which is passed as the second argument to the `error` function. This is similar to the use of `error` we saw in Section 3.3. However, here we also have an extra argument. The first argument is a unique string which identifies the location and nature of the error call: although not required, it is a good idea for each call to `error` to have a unique identifier. In the code above, if a file with the given name does not exist, then execution will transfer to the `error` function and MATLAB will return and halt processing the script and any that have called it. Any further code below this point will not be run. ■

■ Example 4.8

O4.E

In the above example, we might however be a little less strict and decide that if the log file is not present we should still carry out the processing but warn the user that there will be no logging carried out. This is where the `warning` function is useful: the above code can be modified slightly to do this:

```
function processFile(logFileName)

if ( ~ exist(logFileName, 'file') )
  msg = sprintf('No such file: %s. No logging.\n', ...
              logFileName);
  warning('MATLAB:processFile:fileMissing', msg)
end

% further code below
```

In this case, if the log file does not exist, a warning message is displayed but the code below the warning is still executed (although in this example without logging). ■

In other words, both the `error` and `warning` functions can be used to report a problem to the user but the `error` function then returns and stops running the script or function file.

4.6.2 The `try` and `catch` Method

The `error` and `warning` functions described above can be used when we anticipate a particular type of problem that can lead to an error. Sometimes we may want to prepare for a possible error even if do not know in advance what its type will be.

One phrase to describe a program encountering an error is to say that the program has *raised an exception*. The exception itself can be raised for a number of reasons, which include:

- Trying to access an element of an array that is beyond the last element it contains.
- Trying to open a file that does not exist.
- Performing a type error such as trying to perform an arithmetic operation on a variable that is not numeric.
- etc.

If we have a section of code that might raise an exception, we can wrap this code inside a *try–catch clause*. This is a little like an *if–then* clause where a section of code is 'tried' and, if it raises an exception, this exception is then 'caught' by the second part of the clause.

■ Example 4.9

As a simple and artificial example of the use of a try–catch clause, the following is adapted from the MATLAB documentation

O4.E

```
A = rand(3);
B = ones(5);

try
   C = [A; B];
catch

   % handle the error
   msg = ['Dimension mismatch: First argument has ', ...
          num2str(size(A,2)), ' columns, second has ', ...
          num2str(size(B,2)), ' columns.'];
   error('MATLAB:try_catch_ex:dimensions', msg);

end  % end try/catch
```

The code within the `try` section tries to concatenate two arrays vertically, placing `A` above `B`, and then store the result in a third array, `C`. If there is an error when doing this, then the execution of the program will immediately switch to the code inside the `catch` section. ■

It is important to note that *any* error in the `try` section will lead to the code in the `catch` clause being run. This allows different possible reasons for the error to be tested inside the `catch` part of the clause.

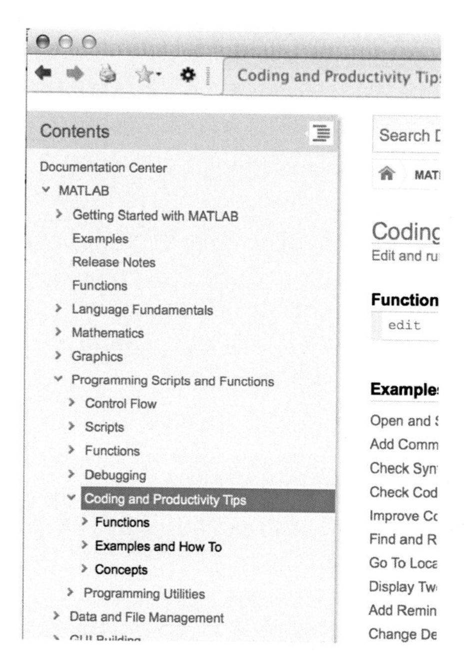

FIGURE 4.3 Accessing the MATLAB help on coding and productivity tips.

4.7 SUMMARY

This chapter has considered how we can incrementally develop a function or set of functions that can be used to solve a programming problem. We have looked at how it is important to make small steps, ensuring that the code runs at each stage and testing its output to see if it behaves as we expect. We have also looked at common types of error such as run-time errors, syntax errors and logic errors and what we can try and do to identify and fix them, i.e. to *debug* them. We have also briefly looked at how to handle errors in our code, testing for them explicitly or using try–catch clauses to wrap sections of code that might raise an error.

4.8 FURTHER RESOURCES

- The MATLAB help is always a good place to start. From the main help menu for MATLAB, go to the section on *Programming Scripts and Functions* (see screenshot in Fig. 4.3) and browse the subsections. For example, the part on *Coding and Productivity Tips* lists various examples and how-to documents (see Fig. 4.3).
- There is a very good MATLAB coding style set of guidelines by Richard Johnson at: http://www.datatool.com/downloads/matlab_style_guidelines.pdf.

- A nice paper on good practice when programming, with a more scientific perspective, can be found at the link: journals.plos.org/plosbiology/article?id=10.1371/journal.pbio.1001745.

EXERCISES

■ Exercise 4.1

O4.C, O4.D

The following code is intended to loop over an array and display each number it contains one at a time. It is quite badly written and needs to be corrected.

```
1  % Display elements of an array one by one.
2
3  % Array of values
4  myArray = [12, -1, 17, 0.5]
5
6  for myArray
7    disp(myArray)
8  end
```

1. Type the code into the MATLAB editor, and use the hints from the editor to locate any syntax errors. While there are syntax errors in the code, the MATLAB editor will not allow you to use breakpoints for debugging. Once you have corrected any syntax errors, insert a breakpoint at the start and run it. Step through the code to identify further possible errors and fix them as you go along.
2. Identify and fix any style violations in the code, i.e. things that do not prevent the code from running but can be changed to improve the appearance or behavior of the code. ■

■ Exercise 4.2

O4.C, O4.D

Type the following code into the MATLAB editor (note that it contains errors).

```
clear, close all

% Get 30 random integers between -10 and 10
vals = randi(21, 1, 30) - 11;

for k = 1:1:numel(vals)
  if (vals(n) > 0)
    sumOfPositives = sumOfPositives + vals(n)
  end
end

disp('The sum of the positive values is:')
disp(sumOfPositives)
```

1. Identify and explain what the different parts of this piece of code are trying to do. What seems to be the main aim of the code?
2. Find the errors and style violations in the code and fix them using the debugger. ■

■ Exercise 4.3

O4.C, O4.D, O4.E

A function with errors in it is given below. It aims to evaluate a quadratic polynomial for a given x value or for an array of given x values. Type this function into a file and save it with the correct name.

```
function [ ys ] = evaluate_quadratic(coeffs, xs)
% Usage: evaluate_quadratic(coeffs, xs)
%
% Find the result of calculating a quadratic
% polynomial with coefficients given at the x
% value(s) given. Return the result in the
% variable ys which must contain the same
% number of elements as xs.

clear
close all

ys = zeros(1, xs);
ys = a * xs^2 + b * xs + c;

end
```

Now type the following into a script, and call it *testScript.m*. This will be used to test the function above.

```
% Test script for evaluate_quadratic.m
clear, close all

x = 3;
coeffs = [1 2 1];
y = evaluate_quadratic(coeffs, x);
fprintf('f(%0.1f) = %0.1f\n', x, y)

xs = -1:0.5:3;
ys = evaluate_quadratic(coeffs, xs);
disp(xs), disp(ys)
```

1. Try to run the script with the debugger, placing breakpoints in the main function. List the errors in the function and suggest how they should be fixed.
2. Add code to the corrected function to validate the input arguments `coeffs` and `xs` to make sure that they can be correctly processed by the function. ■

The following exercises contain longer, more open-ended problem solving activities. There are many ways that they can be tackled and you are encouraged to adopt the incremental programming style described in the chapter. Try to break the tasks of each problem down into multiple functions, these could perhaps be called by a single main function or script. Use the debugger to check for errors as you go along and try to make use of the MATLAB error handling functions, such as `error` and `warning`, to guard against misuse of the code.

■ Exercise 4.4

O4.A, O4.B, O4.C, O4.D, O4.E

An *oscillatory* monitoring device automatically measures systolic and diastolic blood pressure as well as heart rate. The device works by placing a cuff on the arm and inflating it to a pressure that blocks off blood flow through the artery. The pressure is gradually reduced until blood starts to flow again – the pressure at which this happens is the patient's systolic blood pressure. After the flow restarts, the pressure exerted by the cuff causes a vibration in the vessel walls which can be detected by the device. The vibrations stop when the pressure falls below the patient's diastolic blood pressure. The vibrations are *oscillatory* (hence the name of the device) and the frequency of these oscillations gives an estimate of the patient's heart rate.

The file *blood_pressure.mat* contains a set of the device's measurements. Specifically:

- `p`: an array of pressure measurements (without the oscillations) in mmHg, which gradually decreases throughout the monitoring.
- `peaks`: an array of ones and zeros, in which a one indicates a peak in the oscillatory vibrations.

Both arrays are measured with a temporal resolution of 0.05 seconds (i.e. 20 measurements per second).

Write a MATLAB function *m*-file that takes these two arrays as input arguments and returns three values as the output of the function: the systolic and diastolic blood pressures and the heart rate. Write a separate short script *m*-file that reads in the data provided, calls the function and displays the results.

To summarize the key points:

- The systolic blood pressure is the value of `p` at the index where the first non-zero value of `peaks` occurs.
- The diastolic blood pressure is the value of `p` at the index where the last non-zero value of `peaks` occurs.
- The heart rate is the number of heart beats per minute. One way this can be calculated is by first estimating the average time of a single heart cycle, i.e. the average time between non-zero values in the array `peaks`. ■

■ Exercise 4.5

O4.A, O4.B, O4.C, O4.D, O4.E

Exercise 3.9 made use of data acquired with a respiratory bellows. In this exercise we analyze these data further.

The file *bellows_sub.mat* contains a subset of the original bellows signal. It contains the data acquired when the subject was taking deep breaths. The goal of this exercise is to automatically determine the average length of a respiratory cycle from this subset of the data.

The program should perform the following tasks:

- Load in the bellows signal data from the file.
- Determine the measurement times (in seconds) of the bellows data. To do this, use the fact that the bellows signals were acquired at a rate of 500 Hz. You can assume that the measurements start at a time of zero seconds.
- Smooth the signal data to eliminate peaks that are caused by noise. You can use the MATLAB `smooth` function for this purpose (which is part of the curve fitting toolbox), and you will need to experiment with different levels of smoothing, which you can specify by changing the value of the second argument to `smooth` (the *span* of the averaging that is performed). See the MATLAB documentation for `smooth` for details.
- Produce a plot with the smoothed bellows signal on the *y*-axis and the measurement time on the *x*-axis. This plot will help you judge whether the smoothing level you chose in the previous step was sufficient.
- Find the measurement times of all *peaks* in the smoothed bellows signal. Recall that you wrote a function to perform this task in Exercise 3.6, so it would make sense to reuse the code you wrote then.
- Find the lengths of all *respiratory cycles* (i.e. the times between successive peaks) and compute the average of these times.

Your final program should display the plot of the smoothed bellows data, and print to the command window the times of all peaks and the average length of the respiratory cycles. ■

■ Exercise 4.6

O4.A, O4.B, O4.C, O4.D, O4.E

A research project wants to produce a portable monitoring device for measuring blood sugar levels. Normal blood sugar levels for adults are 4.0–5.9 mmol/L, but may rise to 7.8 mmol/L after meals. A level that is higher than 7.8 mmol/L may indicate diabetes.

The data files called *sugar1.txt, sugar2.txt, ... sugar5.txt* contain measurements made by a prototype of the device. A program is required that reads in a single file and:

- Reports the times and blood sugar levels of any measurements that were above 7.8 mmol/L or below 4.0 mmol/L.

- Reports the times and average blood sugar levels for any given 2 hour period in which the average blood sugar was greater than 5.9 mmol/L.

The measurements in the files were taken every 10 minutes over a period of 24 hours, i.e. there are 144 measurements in each file.

Develop a program to meet these requirements. ■

■ Exercise 4.7

O4.A, O4.B, O4.C, O4.D, O4.E

Monte Carlo simulations are a class of algorithms that use repeated random sampling to obtain numerical results. They have found practical use in a range of biomedical applications [5].

This exercise is about using a Monte Carlo simulation to estimate the value of π. The method uses random points chosen inside the unit square in the 2D Cartesian axes, i.e. $0 \leq x \leq 1$ and $0 \leq y \leq 1$. This is illustrated in the figure below which shows part of the unit circle and random points in the unit square.

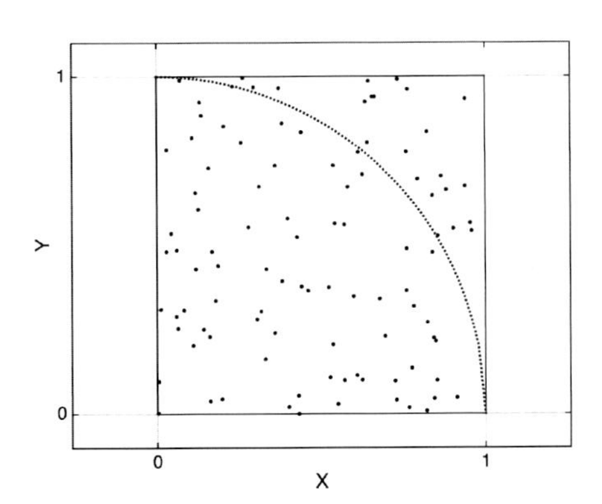

The method picks random points inside the unit square, and these are shown by dots in the diagram. Some of these points will lie inside the unit circle (which is also plotted) and some of the points will lie outside the unit circle.

We can estimate the area of the quarter circle by finding the fraction of points that are inside it. From this, we can estimate the area of the whole circle and therefore the value of π. For example, if we pick 20 points inside the square and 17 turn out to be also inside the unit circle the estimate for the area of the circle (i.e. π) is $4 \times 17 \div 20 = 3.4$.

Write MATLAB code to do this. Break the problem down into a number of sub-tasks and use different functions to carry out each sub-task. Use a main function or main script to control the flow of the program. The user should be able to choose how many points are picked and the program should print a suitable output message.

Run your program and experiment with different numbers of points to see the effect on the estimate of π. ■

■ Exercise 4.8

O4.A, O4.B, O4.C, O4.D, O4.E

The trapezium rule can be used to estimate the integral of a function between a pair of limits. The formula for the trapezium rule is

$$\int_a^b f(x)dx \approx \frac{\Delta x}{2}(f(x_1) + 2f(x_2) + 2f(x_3) + \ldots + 2f(x_{n-1}) + f(x_n))$$

where f is the function to integrate, a and b are the limits of the integration and $\{x_1, x_2, \ldots, x_{n-1}, x_n\}$ are a sequence of points on the x-axis with $x_1 = a$ and $x_n = b$. Δx is the width of the interval between successive points. See the figure below for an illustration.

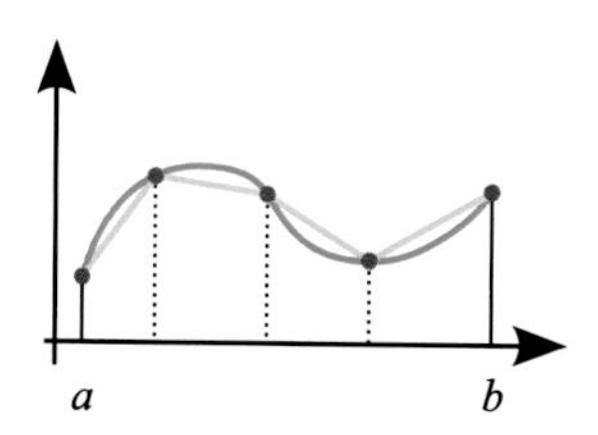

Write code that uses the trapezium rule to estimate the integral of the specific function $f(x) = \sin x$ between a pair of limits and for a given number of points. The user should be able to decide the limits and the number of points, for example they might choose ten points between $x = 0$ and $x = \pi/4$. ■

FAMOUS COMPUTER PROGRAMMER: JOHN VON NEUMANN

John von Neumann was a mathematician who was born in 1903 in Hungary. Von Neumann was a child prodigy. As a 6-year-old, he could divide two 8-digit numbers in his head, and was fluent in both Latin and Ancient Greek. By the age of 8, he was familiar with differential and integral calculus. By the time he was 19, von Neumann had published two renowned mathematical papers. He received his PhD in mathematics at the age of 22.

In 1933, at the age of 30, von Neumann left Hungary and joined Princeton University in the USA. He remained there for his entire career. He was a remarkable academic figure, who made major contributions to a wide range of subjects, including mathematics, physics, economics, statistics and computer science. He was also a principal member of the Manhattan project that developed the world's first atomic weapon.

Von Neumann was one of the founding figures of computer science. He was strongly involved in the early work on developing digital computers (the EDVAC project). In this project, he was credited with proposing a computer architecture in which the program and data are both stored in the same computer memory. In reality, this idea was based on the earlier proposal of Alan Turing (see Chapter 2's Famous Computer Programmer) and von Neumann acknowledged this debt to Turing's work. However, it was von Neumann's contribution that led to the first practical implementation of this concept, which is known today as the 'von Neumann architecture'.

In his personal life, John von Neumann was a sociable figure and was known for throwing large parties whilst at Princeton. However, he was a notoriously bad driver. When he was hired as a consultant at IBM, they had to quietly pay the fines for his multiple traffic offenses. He enjoyed music, and regularly received complaints from colleagues at Princeton (including Albert Einstein) for playing his German marching music excessively loud at the office.

John von Neumann died of cancer in 1957. Because of his involvement in the USA's atomic weapons program, his final days were spent in a military hospital in case he revealed any military secrets on his death-bed. There has been speculation that his cancer was the result of his attendance at some of the early atomic weapons tests.

> *"You wake me up early in the morning to tell me that I'm right? Please wait until I'm wrong."*
>
> **John von Neumann**

CHAPTER 5

Data Types

LEARNING OBJECTIVES

At the end of this chapter you should be able to:

O5.A Explain how a data type acts to help interpret the sequences of 1s and 0s stored in the computer's memory
O5.B Identify the fundamental data types used in MATLAB
O5.C Make appropriate use of integer and floating point numeric types and describe the differences between them
O5.D Use MATLAB commands to identify the type of a variable
O5.E Convert between types and explain when this is done automatically by MATLAB
O5.F Use cell arrays to contain strings and other data types more flexibly
O5.G Explain how different types of array can be created and accessed using different types of bracket: [ ], ( ) and { }
O5.H Make use of the `struct` data type in MATLAB

5.1 INTRODUCTION

As we have already seen, MATLAB easily handles numeric data and array manipulation is one of its strengths. As discussed in Section 1.5, MATLAB can also process other, non-numeric data types and these can also be collected in arrays.

Data types in MATLAB can be broadly divided into fundamental types for everyday use and more advanced types. The fundamental types include the basic data types discussed in Section 1.5 (numeric values, characters, and Boolean values) as well as arrays constructed of these types. More advanced types include cells and structures as well as maps, tables and handles.

Ultimately, any data in a computer's memory is stored as a sequence of 1s and 0s. These are the *bits* used to represent the data. A *byte* is a sequence of eight bits. Data are normally stored using a number of bytes, i.e. in multiples of 8 bits.

MATLAB Programming for Biomedical Engineers and Scientists. DOI: 10.1016/B978-0-12-812203-7.00005-7

A *data type* represents a way of interpreting a sequence of bits in terms of some higher level representation. A pattern of 1s and 0s arranged over a sequence of bytes can represent different things *depending on how they are interpreted* and this underlies what it means to have different data types.

■ Example 5.1

O5.A

Any sequence of bits (in a byte) can be interpreted as an integer. For example, the sequence 01000001 can be viewed as the number 65 because 01000001 is the *binary* representation of 65 (i.e. $2^6 + 2^0 = 64 + 1$).

We can, however, also interpret the same pattern as the character 'A'. This is because of the definitions given in the ASCII[1] code, which defines a correspondence between the number 65 and the character 'A'. The correspondence continues in the alphabet, for example 'B' matches with the same bit pattern that would be used for 66, 'C' matches with 67 etc. ■

■ Activity 5.1

O5.A

How would the letter 'D' be represented as a binary number in the computer's memory? ■

5.2 NUMERIC TYPES

We briefly discussed the fundamental data types offered by MATLAB, including numeric types, in Section 1.5. To recap, we distinguish between two categories of numeric types: those used to represent integers and those used to represent floating point numbers (i.e. numbers with a fractional part). Types used for integers can be *signed* (these are allowed to be positive or negative as well as zero) or *unsigned* (for non-negative numbers only).

Different integer data types can use different numbers of bytes to store their values. Using more bytes, it is possible to represent larger integers. The number of bytes used to store a floating point number can also vary. If more bytes are used, as well as being able to represent larger floating point numbers, we can also represent the fractional part with greater *precision*.

The names of the MATLAB types that are used for integers are given below.

- `uint8`, `uint16`, `uint32`, `uint64`
- `int8`, `int16`, `int32`, `int64`

The different types are prefixed with a `u` if they are unsigned. The number at the end indicates how many bits each one uses: 8, 16, 32 or 64, corresponding to 1, 2, 4 or 8 bytes respectively.

---

[1] American Standard Code for Information Interchange.

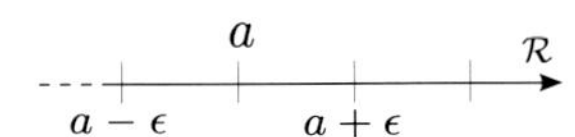

FIGURE 5.1 Illustration of numeric value precision.

The types that MATLAB uses to represent non-integer numbers are called `single` and `double`, meaning *single precision floating point numbers* and *double precision floating point numbers* respectively. Again, these use different numbers of bytes. See below for further details on the precision of floating point numbers.

5.2.1 Precision for Non-Integer (Floating Point) Numeric Types

The precision of a floating point number relates to the smallest distinguishable decimal place that can be used within it. In a computer's memory, precision can be represented by the difference between a number that it is possible to represent and the *next nearest* representable number (above or below). For example, in Fig. 5.1, a value a that is stored in memory would have two adjacent representable numbers $a - \epsilon$ and $a + \epsilon$ on the real line, $\mathcal{R}$. For high precision, the value of ϵ is very small, i.e. adjacent representable floating point numbers are very close. Because of the way floating point numbers are represented in memory, *the size of* ϵ *depends on the size of the number* a. For a large value of a, there will be a larger gap to the nearest representable numbers than there would be if the value of a were small.

The built-in MATLAB function `eps` (short for epsilon) can be used to find out the value of ϵ for a given number and data type. If we call `eps` with the name of the data type as an argument, it will return the value of ϵ for a value of $a = 1.0$ when represented in that data type.

```
>> eps('double')
ans =
   2.2204e-16

>> eps('single')
ans =
  1.1921e-07
```

In each case, we obtain the gap between $a = 1.0$ and the adjacent representable numbers when we use `double` or `single` data types.

If we call the `eps` function with a number as its argument, it will return the value of ϵ for that number when using the `double` type. For example

```
>> eps(1000)
ans =
   1.1369e-13
```

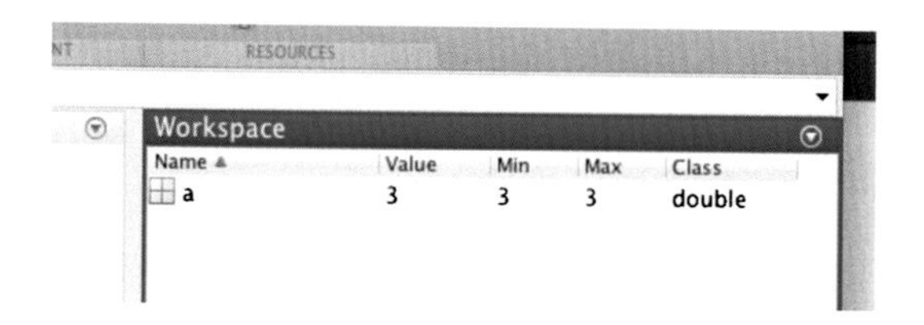

FIGURE 5.2 The workspace browser shows information about the type of a variable.

shows that the gap between 1000 (stored as a `double`) and the next representable number is 1.1×10^{-13} which is significantly larger than the gap for 1.0 (2.2×10^{-16}, as calculated earlier).

5.2.2 MATLAB Defaults to Double Precision for Numbers

An important point to remember when using numeric data is that MATLAB always uses double precision for numeric data by default. For example, typing `a = 3` at the command window followed by a call to the `whos` function gives the following output:

```
>> whos
  Name      Size            Bytes  Class     Attributes

  a         1x1                 8  double
```

We can see that there is only one variable, `a`, in our workspace and the `whos` function tells us that

- MATLAB has used the `double` data type to represent our variable (under the heading `Class`).
- MATLAB views our variable implicitly as an array – with a size of 1×1. Even a scalar value is considered in MATLAB to be a (small) array.
- The amount of memory used to represent the variable is 8 bytes (i.e. 64 bits).

Similar information is shown by the workspace browser in the MATLAB environment, as shown in Fig. 5.2. Note that we can choose the columns displayed in the workspace browser by clicking the small triangle in its top bar to open its context menu.

5.2.3 How Does MATLAB Display Numeric Values by Default?

We need to take some care when inspecting how MATLAB displays numbers in the command window as this can vary depending on the size of the number. By default, MATLAB tries to display numbers as compactly as possible in what is known as a 'short' format. For example, asking for the value of π in the command window gives

```
>> pi
ans =
    3.1416
```

This does not mean that MATLAB stores only the first four decimal places of π, it is simply only *showing* the value rounded to 4 decimal places.

We can request that MATLAB uses a longer format to display numbers by typing `format('long')` in the command window.

```
>> format('long')
>> pi
ans =
   3.141592653589793
```

We can revert to the default format by typing `format('short')`.

For large numbers, MATLAB uses a compact scientific notation. In scientific notation, a number is written in the form $A \times 10^B$ where A is called the *mantissa* and B is called the *exponent*.

For example, the value of the factorial of 15 is around 130 billion (to be exact $15! = 130{,}767{,}436{,}800$). However, the default way in which this number is displayed in the MATLAB command window is as follows:

```
>> factorial(15)
ans =
   1.3077e+12
```

Which is 1.3077×10^{12}, which means that MATLAB has again rounded to 4 decimal places (this time for the mantissa) when displaying the result. The exact value in scientific notation would be $1.307674368 \times 10^{12}$.

This way of displaying numeric values can require some care when looking at arrays, especially if an array contains both small and large values. For example, let's initialize an array using three integers with very different sizes

```
>> x = [25 357346 2013826793]
x =
   1.0e+09 *

    0.0000    0.0004    2.0138
```

We can see that MATLAB tries to show all three numbers in scientific notation using *a single exponent*. In this case the exponent is 9 and this is determined by the largest number in the array. If we were to write all three numbers *exactly* using this exponent, then the three numbers would appear as:

0.000000025×10^9 $\quad$ 0.000357346×10^9 $\quad$ 2.013826793×10^9

MATLAB shortens this to a single array with a common exponent:

$10^9 \times [0.000000025 \quad 0.000357346 \quad 2.013826793]$

Finally, applying short notation to each of the mantissa values in the array means rounding them to four decimal places. This is why the array appears in MATLAB as

```
x =
   1.0e+09 *
    0.0000     0.0004     2.0138
```

This does not say that the first number in the array is zero. It just means that the use of a large exponent has meant that the non-zero decimals in the mantissa have 'gone beyond' the 4th decimal place.

5.2.4 Take Care when Working with Numeric Types Other than Doubles

MATLAB's default behavior of using the `double` data type for numeric data is fine most of the time, even if, for example, we only want to work with integers. *The best advice is to leave the default behavior.*

However, this default behavior can be a problem on occasion, for example if we want to work with *very* large amounts of data. In such cases, we may need to explicitly force MATLAB to use one of the integer types above, or the `single` type if we still need to work with floating point data. This is because these types use fewer bytes so our data will require less computer memory.

■ Example 5.2

O5.B, O5.C

Here, we create array variables of the same length but we force MATLAB to create them using a number of different data types as well as the default `double` type.

```
>> a = zeros(100000,1);
>> b = int8(zeros(100000,1));
>> c = int16(zeros(100000,1));
>> whos
  Name        Size               Bytes  Class    Attributes
  a           100000x1           800000  double
  b           100000x1           100000  int8
  c           100000x1           200000  int16
```

Note the different amounts of memory required to store each variable. ■

One warning here is that variables which have a type other than `double` will no longer work properly with functions that expect a `double` as their argument(s), and there are quite a few of these. For example, a call to the square root function such as the following will now generate an error:

```
>> a = int16(3);
>> b = sqrt(a);

Undefined function 'sqrt' for input arguments of type 'int16'.
```

whereas if we had simply set `a=3`, the call to `sqrt` would have worked because the variable `a` would have been, by default, a `double` precision floating point number and therefore suitable for passing to the `sqrt` function.

5.2.5 Ranges of Numeric Types

Another difficulty that can be encountered when not using MATLAB's default `double` type for numbers relates to the range of values that a numeric type can store.

We start by looking at the ranges of some of the integer data types.

■ Example 5.3

For example, an unsigned single byte integer (`uint8`) can only store values in the range 0–255. We can check this using the built-in functions `intmax` and `intmin`, which take a string specifying the data type.

O5.C

```
>> intmax('uint8')
ans =
  255
>> intmin('uint8')
ans =
    0
```

This tells us that if we use eight bits to represent an unsigned integer, then we can represent the numbers from 0 to 255 inclusive. The number 255 is the decimal value that corresponds to the binary number 11111111, i.e. the one where all the eight bits are set to 1.

If we use 16 bits, then we can represent a larger range of unsigned integers, for example

```
>> [ intmin('uint16') , intmax('uint16') ]
       0  65535
```

Note how the range changes if we want to represent *signed* integers, i.e. both positive and negative numbers:

```
>> [ intmin('int16') , intmax('int16') ]
 -32768  32767
```

■

We need to take care when working with mixed numeric types because we can have unexpected behavior if we try to represent a number beyond the range of a data type we are using. This is shown by the following command window calls:

```
>> a = 250 + 10
a =  260

>> b = uint8(250 + 10)
b =  255

>> whos
  Name       Size            Bytes  Class     Attributes
  a          1x1                 8  double
  b          1x1                 1  uint8
```

Both commands aim to set the variables to a value of 260, but this failed in the case of variable `b`, which was forced to be a one byte unsigned integer. Therefore, the assignment to `b` gave an incorrect result due to what is called an *overflow* error.

To re-iterate, if we leave all numeric values in MATLAB's default data type of `double` then we reduce the chance of such unexpected behavior.

■ Example 5.4

O5.C

We can check the range of *floating point* numbers using the `realmax` and `realmin` functions.

```
>> realmax('single')
ans =
  3.4028e+38

>> realmax('double')
ans =
  1.7977e+308
```

This shows that the biggest `single` precision number we can represent is 3.4028×10^{38} whilst the biggest `double` precision number is 1.7977×10^{308}. ■

■ Activity 5.2

O5.A, O5.B, O5.C

Find the problems with the following pieces of code and rectify them.

1. Code to compute and display the value of e^8:

```
% Compute value of e^8
x = int16(8);
y = exp(x)
```

2. Code to compute and display the exponential sequence 1, 10, 100, 1000, etc.:

```
% Generate exponential sequence 1,10,100,1000, ...
a = 1:10;
b = 10^a;
disp(b);
```

3. Code to compute the squares of an array of numbers:

```
% Compute squares of array
w = uint8([1 16 24 19 21 12]);
q = w.^2
```

■

5.3 INFINITY AND NAN (NOT A NUMBER)

Two special MATLAB identifiers are used to deal with cases where we encounter infinities of numbers or numbers that cannot be determined.

The first of these is `inf`. MATLAB uses `Inf` to represent infinite numbers. For example:

```
>> 1/0

ans =
   Inf
```

The function `isinf` can be used to test if a number is infinite. It can be applied to an array of numbers and will return 1 (`true`) at locations with an infinite value.

■ Example 5.5

For example, consider the following element-wise division for two arrays: O5.C

```
>> a = [1 2 3];
>> b = [2 0 1];
>> a ./ b

ans =
    0.5000         Inf    3.0000

>> isinf( a ./ b )
ans =
     0     1     0
```

The 1 in the second element of the result indicates that `isinf` returned a `true` value for the result of the operation `2/0`. ■

Another special MATLAB identifier is `NaN`. This is used to represent indeterminate values or values that cannot be interpreted as a number. For example, the expression `0 / 0` cannot be interpreted as a number so is represented by

NaN. NaN is returned by functions that attempt to interpret strings as numbers when the given string cannot be interpreted properly. The function isnan can be used to test for NaN values in a similar way to the isinf function and it can also be applied to arrays.

■ Example 5.6

O5.C

For example, consider the following calls to the built-in str2double function:

```
>> str2double('2.3')
ans =
    2.3000

>> str2double('i')
ans =
   0.0000 + 1.0000i

>> str2double('pi')
ans =
   NaN

>> n = str2double('blah')
n =
   NaN

>> isnan(n)
ans =
   1
```

■

■ Activity 5.3

O5.C

The *end diastolic volume* (EDV) and *end systolic volume* (ESV) of the left ventricle of the heart refer to the volume (in mL) of blood just prior to the heart beat and at the end of the heart beat respectively. The *ejection fraction* (EF) is the fraction of blood pumped out from the heart in each beat, and can be computed from the EDV and ESV as follows:

$$EF = \frac{EDV - ESV}{EDV}.$$

The file *edv_esv.mat* contains EDV and ESV values for a number of patients. Both are integer values. Write a MATLAB script *m*-file to read in these data and compute EF values for each patient. The script should then display the EDV, ESV and EF values for all patients to the command window. Note that the data may contain some erroneous values so you should write your code in such a way that it can detect such cases and display a message indicating that there is a problem with the data. ■

5.4 CHARACTERS AND STRINGS

It is often necessary to use non-numeric types. Important examples are characters and strings, which are often used to report output or communicate with a user. In MATLAB, single quotes are needed to enclose strings and characters. We can assign a single character to a variable with a command such as `s = 'A'`. The command `whos` allows us to inspect how MATLAB represents this variable:

```
>> whos
  Name       Size             Bytes  Class    Attributes
  s          1x1                  2  char
```

This shows us that MATLAB views the type as a size 1×1 array with a single element of the character type (i.e. its `Class` is `char`). It also shows us that two bytes are needed to store this single character.

We can also assign a string of characters in exactly the same way. For example, typing `s = 'ABC'` followed by `whos` gives:

```
>> whos
  Name       Size             Bytes  Class    Attributes
  s          1x3                  6  char
```

As we can see, we still have an array with elements of type `char` but now its size is 1×3 and it uses 6 bytes of memory (2 for each element).

The above illustrates that a string is represented by MATLAB as an array of character elements. We can make this explicit during assignment by using the command `s=['A', 'B', 'C']` which achieves exactly the same result as `s='ABC'`.

Generally, strings are arrays of characters with size $1 \times N$ where N is the number of characters in the string. Therefore, they can be viewed as row vectors and we can concatenate two or more strings to generate a longer array.

■ Example 5.7

For example, the following commands create and manipulate some string variables. *O5.B*

```
>> s1 = 'hello';
>> s2 = ' ';
>> s3 = 'world!';
>> s = [s1, s2, s3]

s =
hello world!

>> size(s)
```

```
ans =
     1    12
```

Concatenating strings row-wise in this way produces a single long string with twelve elements (characters). ■

If we want an array that contains three *separate* strings (as opposed to a single long string) then this is more difficult. One way to do it is to concatenate the strings column-wise (using the semi-colon to separate entries). Doing this can, however, lead to an error. For example, using `s1`, `s2` and `s3` from the previous example, we would get the following:

```
>> s = [s1 ; s2 ; s3]

Error using vertcat
Dimensions of matrices being concatenated are not consistent.
```

This is because the lengths of the strings `s1`, `s2` and `s3` are 5, 1 and 6 respectively. MATLAB is trying to place the three row vectors, one beneath the other, to create a 2-D array. This is only possible if all of the row vectors have the same length (i.e. they must be consistent). We can make the strings consistent by modifying them.

■ Example 5.8

O5.B

For example, the following commands address this issue by *padding* some of the strings with spaces so that they all end up having the same length.

```
>> s1 = 'hello ';
>> s2 = '      ';
>> s3 = 'world!';
>> s = [s1 ; s2 ; s3]
s =
hello

world!

>> size(s)
ans =
     3     6
```

The extra spaces in `s1` and `s2` mean that they now each have a length of 6, the same as `s3`. We concatenate the three row vectors vertically, to create a 3×6 array of characters. ■

However, this approach is quite cumbersome and a better way to have an array of strings *and* to keep them separate is by using a *cell array* (see Section 5.7). A cell array can act as a container array for variables with different types.

■ Activity 5.4

The file *patient_names.mat* contains a 2-D array of characters. Each row of the array is a string of length 20 characters representing a patient name. Each patient name consists of a first name, followed by a space, then a second name and then more spaces to make its length equal to 20. Write a program to read in this array, separate each name into a first and second name and then build up two new 2-D arrays of the first and second names respectively. These two new arrays should be written to new *MAT* files. The new 2-D arrays of first and second names that you produce should also be padded with spaces so that each name is a string of 20 characters.

O5.B

The MATLAB functions `strfind` and `blanks` might be useful to you. Look at the MATLAB documentation for details of how to use them. ■

5.5 IDENTIFYING THE TYPE OF A VARIABLE

Before we look at further data types, we will briefly consider how we can identify the types of variables in MATLAB.

As we have seen in Section 1.5, we can use the `whos` command to display a summary description of the variables that are in the workspace. This is useful, but sometimes we may have many variables in the workspace and we may not need to know about some of them. Also, the `whos` command gives information on the size of variables (in bytes) which we may not want.

The built in function `class` can also be used to identify the type of a variable. It produces a string that describes the variable type.[2]

■ Example 5.9

For example, consider the following sequence of commands.

O5.B, O5.D

```
>> a = 1:5
a =
     1     2     3     4     5

>> b = 'Some String'
b =
Some String

>> class(a)
ans =
double

>> class(b)
```

[2] Note: The use of the word 'class' in MATLAB to describe the data type of a variable is at odds with the use of the same word in other programming languages, such as C++ or Python, where the word 'class' is reserved for something much more specific.

```
ans =
char
```

Here, we can see that the `class` command has identified the types of the variables `a` and `b` as `double` and `char` respectively. ■

Another way to identify the type of a variable is to use the built-in `isa` function. This function takes a variable and a string as its arguments (where the string describes a type) and returns a Boolean result (see Section 5.6), with a 1 indicating `true` (the variable *is* of the specified type) and a 0 indicating `false` (it isn't).

Further functions for testing data types are specific to each type and do not need a string description of the type to be given. For example, we can test if a variable is a character or a character array with the function `ischar`, and the function `isnumeric` can be used to test if a variable's data type is one of the numeric types (i.e. `uint8`, `int64`, etc.).

■ Example 5.10

O5.B, O5.D

This example uses the same variables defined in the previous example.

```
>> isa(a, 'double')
ans =
     1

>> isa(a, 'char')
ans =
     0

>> ischar(b)
ans =
     1

>> isnumeric(b)
ans =
     0
```

■

5.6 THE BOOLEAN DATA TYPE

In programming, we use the term 'Boolean' to refer to data and operations that work with the two values 'true' and 'false'. As described in Section 1.5, MATLAB uses the `logical` data type for this. For this type, MATLAB can also use the identifiers `true` and `false`, representing them internally with the numbers 1 and 0 respectively.

For example, here we assign a value of `true` to a variable, display it and ask MATLAB what class it is:

```
>> v = true;
>> disp(v)
     1

>> class(v)
ans =
logical
```

Note that when the value is displayed, it is shown as the number 1 even though the data type is `logical` (i.e. Boolean).

A common way to generate logical *arrays* is with comparison operators. Here, we generate a list of random numbers between 0 and 1 and test which elements are greater than 0.5:

```
>> a = rand([1,5])
a =
    0.1576    0.9706    0.9572    0.4854    0.8003

>> b = a > 0.5
b =
     0     1     1     0     1
```

The `whos` command can give a description of these variables, where we see that the data type of variable `b` is logical:

```
>> whos
  Name      Size            Bytes  Class      Attributes

  a         1x5                40  double
  b         1x5                 5  logical
```

■ Activity 5.5

Consider the code listing shown below, which defines a number of different variables to store data about a hospital patient. *O5.B, O5.D*

```
name = 'Joe Bloggs';
initial1 = name(1);
initial2 = name(strfind(name, ' ')+1);

heart_rates = [76 80 95 110 84 70];
cholesterol = 4.3;

bradycardia = heart_rates < 60
tachycardia = heart_rates > 100

class(name)
ischar(initial1)
isnumeric(initial2)
isa(heart_rates, 'int16')
class(cholesterol)
```

```
class(bradycardia)
class(tachycardia)
```

Predict what you think the output of the code would be. Enter the code into a MATLAB script *m*-file, run it and check if your predictions are correct. ■

5.7 CELLS AND CELL ARRAYS

As we saw in Section 5.4, it would be useful to be able to easily store a number of strings with varying lengths in a single array. This can be done using an array containing a MATLAB-specific data type known as a *cell*.

We can construct a cell array using curly brackets: '{' and '}'. Recall that we normally use the square brackets '[' and ']' to construct an ordinary array (e.g. see Example 5.7). Here, we initialize a cell array to contain three strings with different lengths.

```
>> a = {'Hi', ' ', 'World!'}
```

A call to `whos` gives some details of what is stored:

```
>> whos
  Name      Size            Bytes  Class    Attributes

  a         1x3               354  cell
```

We say that the array contains three *cells* and each cell contains a string. The strings themselves are 2, 1 and 6 characters long. Putting these strings together into a cell array, we see that it takes up 354 bytes in memory – much more than we would expect for three short strings. Cell arrays are therefore not efficient in terms of memory usage, but as long as we are not seeking to store too much data, the convenience of a cell array outweighs the extra memory needed.

There are two ways in which we can now access the data in the cell array. One is where we view its elements as cells or sub arrays of cells. The other way is where we view the elements in terms of their 'native' data type – which in our example was the string data type (i.e. a `char` array).

■ Example 5.11

O5.D, O5.F, O5.G

This example illustrates the different ways of accessing the elements of a cell array. We start by putting three strings into a cell array called `names` which is constructed using curly brackets.

```
>> names = {'Bloggs', 'San', 'Andersen'};
```

After constructing the array, the different ways of accessing the data in `names` are illustrated below.

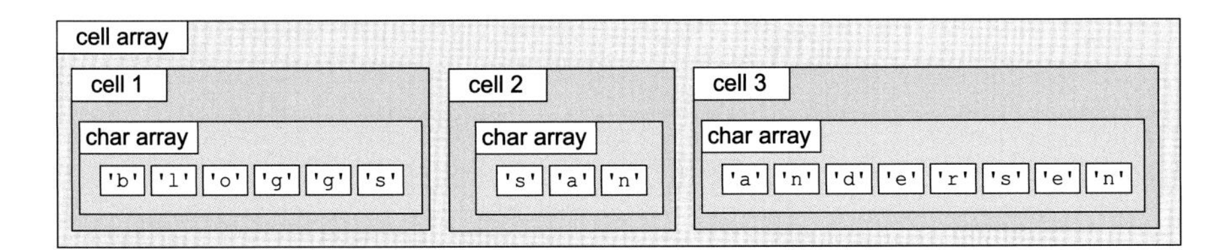

FIGURE 5.3 An illustration of a cell array.

```
>> first_cell = names(1);
>> rest_of_cells = names(2:3);
>> third_string = names{3};
```

In the above code, the cell array created can be visualized by the diagram shown in Fig. 5.3. This illustrates an array containing three cells and each cell contains, in this case, its own `char` array (i.e. string).

After creating the cell array, the variable `first_cell` is assigned the value of a single cell (the first cell in the array `names`). The variable `rest_of_cells` is assigned with a sub-array of cells from `names` (the second and the third). Both variables `first_cell` and `rest_of_cells` are assigned using the round brackets that MATLAB normally uses for accessing array elements: '(' and ')'.

Variable `third_string` is different. This time, curly brackets, '{' and '}', are used to *access the contents of a cell* (the third one) rather than the cell itself. So `third_string` is assigned a `char` array (i.e. a string). We can check the status and types of our variables using the `whos` command.

```
>> whos
  Name                Size         Bytes  Class     Attributes

  first_cell          1x1            124  cell
  names               1x3            370  cell
  rest_of_cells       1x2            246  cell
  third_string        1x8             16  char
```

Note how `first_cell` and `rest_of_cells` are both cell arrays (with sizes 1×1 and 1×2 respectively), whereas `third_string` is a string of length 8. ■

It is important to be clear about the use of the different types of brackets (round, square and curly). The different types of brackets can be used to do one of two things:

- Create an array.
- Access an element (or elements) in the array.

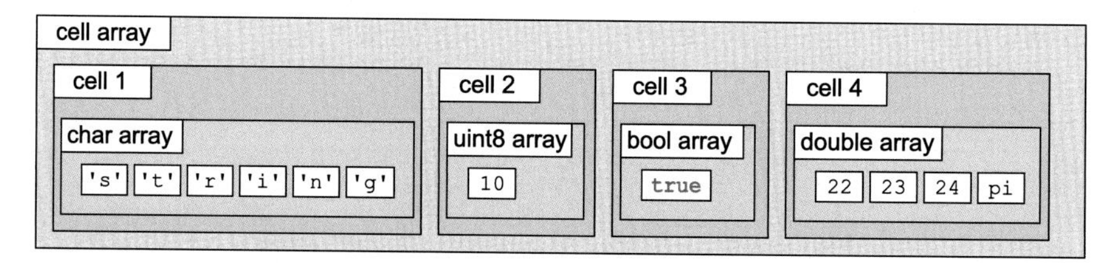

FIGURE 5.4 A cell array containing mixed types.

Not all combinations are possible. For example, in MATLAB we cannot use the square brackets to *access* elements in an array, they are only used for *constructing* arrays. See Section 5.7.2 for more details.

5.7.1 Cell Arrays Can Contain Mixed Data Types

We have seen above how a cell array can be useful for storing a set of strings with varying lengths. In fact, a cell array can be viewed as an indexed list of containers, and each container can hold variables of *any* data type. Here, we define a cell array with a mixture of types:

```
>> mixedData = {'string', uint8(10) , true, [22, 23, 24, pi]}
mixedData =
    'string'    [10]    [1]    [1x4 double]
```

We can use the curly brackets to access the elements inside the cell array and use the `class` function to identify their types:

```
>> class( mixedData{2} )
ans =
uint8

>> class( mixedData{3} )
ans =
logical
```

The `mixedData` cell array is illustrated in Fig. 5.4. The diagram shows how the array contains four cells, each containing an array of a different type.

The ability of a cell array to contain a mixture of data types may be of interest but is rarely useful in practice. Cell arrays are most often used with a uniform data type, i.e. with all of the cells containing data of the same type. We have already seen this in the case of the cell array of strings above.

Another example of a cell array containing a mixture of different data is shown below.

```
>> a = {1, -2, [3,10]}

a =
```

```
    [1]    [-2]    [3,10]
```

In this case, all data are of the same type (`double`), but they are a mixture of single values and arrays. However, we have already seen that MATLAB considers a single value to actually be a 1×1 array.

5.7.2 The Different Kinds of Bracket: Recap

To recap, perhaps the two most two basic operations relating to an array are

- Creating the array.
- Accessing elements in the array.

Depending on whether we have a cell array or a 'standard' array of a basic data type (e.g. `char`), we can distinguish the different behavior of each of the types of bracket: square, curly and round.

Creating an Array

- Square brackets '[ ]':
  Creates an array where all entries have the same type, e.g. `double` or `char`. For example,

  ```
  numArray = [5, 62.7, 3];
  charArray = ['a', 'b', 'c'];
  ```

- Curly brackets '{ }':
  Creates a cell array. Its cells can contain any type. They can contain items of the same type or a mixture of types. For example,

  ```
  sameTypeCells = {'Once', 'upon', 'a', 'time'}
  diffTypeCells = {'Number', 17}
  ```

Accessing Elements

- Round brackets '( )':
  These can be used to access one or more elements from any type of array. The elements returned always have the same type as the array, i.e. for an array of doubles the round bracket will return doubles, for an array of cells they will return cells. For example, using the arrays created above:

  ```
  >> numArray(2:3) % Second and third elements.
  ans =
       62 3

  >> y = sameTypeCells([1, 3]) % first and third
  y =
      'Once'    'a'

  >> class( sameTypeCells(2) ) % Compare with below
  ans = cell
  ```

- Curly brackets '{ }':
  When used for accessing elements, curly brackets are specific to cell arrays. They can be used to obtain the content of a cell *in its native data type*, i.e. not in the form of a cell. For example,

```
>> disp( sameTypeCells{2} )
upon

>> class( sameTypeCells{2} ) % Compare with above
ans = char
```

■ Activity 5.6

O5.D, O5.F, O5.G

Examine the following code, which defines a number of arrays to store patient data.

```
name = 'Joe Bloggs';
heart_rates = [76 80 95 110 84 70];
cholesterol = 4.3;

patient1 = {name, heart_rates, cholesterol};

class(patient1{1})
class(patient1(1))
class(patient1{2})

name = 'Zhang San';
heart_rates = [60 61 58 59 60 66 64];
cholesterol = 7.0;

patient2 = {name, heart_rates, cholesterol};

patients = [patient1 patient2];

class(patients)
class(patients(2))
```

Predict what you think the output of the code would be. Enter the code into a MATLAB script *m*-file, run it and check if your predictions are correct. ■

5.8 CONVERTING BETWEEN TYPES

Now that we have considered some of the fundamental data types, we can look at converting between data types. It can be useful to do this on occasion, for example we may need to read a text file as a set of strings and then re-interpret them as numeric values. The simplest way to convert data types is to use the functions associated with each type.

Common conversions include converting between numbers and characters and converting between numbers and logicals. Some conversions, such as between characters and logicals, are not allowed.

5.8.1 Converting Between a Number and a Character

We stick to the default numeric type `double` in this example. The `char` function can convert a number to a character:

```
>> char(66)

ans =
B
```

To go the other way, i.e. convert a character to a number, we can use the function named `double`[3]

```
>> double('B')

ans =
    66
```

This shows that conversion between numbers and characters is based on the numeric codes, e.g. the ASCII codes, that are used for characters.

5.8.2 Converting Between a Number and a Logical Type

We can convert a number to a logical type using the `logical` function. Consider the following commands.

```
>> logical(1)
ans =
     1
>> logical(3.5)
ans =
     1
>> logical(-12)
ans =
     1
>> logical(0)
ans =
     0
```

These show that the conversion of any non-zero number to a Boolean (`logical`) type will always return `true` (i.e. 1). It will only produce a `false` result if the number is zero (as in the last example above).

Converting a `logical` type to a number can again be carried out using the `double` function. There are of course only two cases to check, and the results are what we would expect. The following commands illustrate the conversion of `logical` values to numbers.

---

[3] Take care to distinguish between the *function* named `double` and the data type `double` that it returns.

```
>> double(true)
ans =
     1

>> double(false)
ans =
     0
```

5.8.3 Converting Arrays

Converting arrays of data is also possible. For example, the function `double` will attempt to convert any argument it is given to a double precision floating point number. If the argument is an array then it should give an array of `double` values. Let's initialize an array `arr` with a list of unsigned 8-bit integers.

```
>> arr = uint8(5:9)
arr =
    5    6    7    8    9

>> class(arr)
ans =
  uint8
```

Passing `arr` as an argument to the `double` function performs a conversion to the `double` type.

```
>> arrTwo = double(arr)
arrTwo =
     5    6    7    8    9

>> class(arrTwo)

ans =
  double
```

Note that we cannot distinguish the arrays above by examining their contents but we can by using the `class` function.

We can convert from numbers to characters using the `char` function. For example,

```
>> arr = [104   101   108   108   111];

>> char(arr)

ans =
  hello
```

In other words, the list of numbers initially put into the array, when converted to characters individually, produces a `char` array containing the characters `'h'`, `'e'`, `'l'`, `'l'`, `'o'`.

MATLAB will sometimes convert values automatically. This can happen, for example, when a mixture of data types are put into an ordinary array.[4] In this case, a set of rules determine how the data types contained in the array are converted so that the resulting array has a uniform type.

■ Example 5.12

This example illustrates automatic type conversion and a possible problem it can cause. We initialize an array as follows:

O5.B, O5.E

```
>> x = 70;
>> str = ['This is number seventy: ' x];
```

Here, we place a string and a numeric type together in the array. MATLAB will automatically try to convert the numeric variable x to a string and concatenate it with the first string. We need to take care as the result may not be what we expected. If the above string is displayed, we get the following:

```
>> disp(str)
This is number seventy: F
```

showing that the conversion was defined by the ASCII codes. The programmer may not have intended this and might instead have wanted to print a string representing the numeric value 70.

MATLAB has a specific function for this kind of conversion, num2str, which can be used as follows:

```
>> y = num2str(x)

y = 70

>> disp(class(x))
double

>> disp(class(y))
char
```

This shows that the variable y contains the characters required to represent the number 70. In particular, it contains two characters: a '7' and a '0'. We can now modify the displayed string above to give a more sensible message:

```
>> str = ['This is number seventy: ' num2str(x)];

>> disp(str)
This is number seventy: 70
```

■

[4] Not a cell array.

Going the other way, the built-in function `str2double` takes a string representation of a number and converts it into a numeric value. For example, compare the different behaviors in the following two command window calls which lead to very different results:

```
>> '3.142' + 1
ans =
    52    47    50    53    51

>> str2double('3.142') + 1
ans =
    4.1420
```

In the first command, the `char` array `['3','.','1','4','2']` was automatically converted to numeric values based on the ASCII codes of each character, then a value of 1 was added to each. In the second command, the string `'3.142'` was re-interpreted (converted) directly as the numeric value that is close to pi (π), and then a value of 1 was added. The second command is perhaps the more sensible one to use in general.

Terminology: Casting. The process of converting from one data type to another is often called *casting*. We say that a variable is cast from one type to another. This cast can be *explicit*, using a function, or it can be *implicit* (i.e. automatic).

Terminology: Strong and Weak Types. Different programming languages vary in how strict they are regarding the type of a variable. Some languages, such as C++, require the type of a variable to be explicitly stated when it is defined. For example,

```
double x = 3.14;
```

Also, a language may allow little or no conversion of the type of a variable. In this case we say that the language is *strongly typed*.

In the examples we have looked at, we can see that MATLAB is fairly permissive about converting types of variables, with many conversions taking place implicitly (automatically). Therefore, we say that MATLAB is a *weakly typed language*.

Good Advice: In general, it is always safer to convert between data types using explicit function calls (e.g. `double`, `char`, `str2double`, etc.) rather than relying on automatic conversions. This leads to clearer code that is easier to debug.

■ Activity 5.7

O5.B, O5.E

In Example 1.2 we wrote code to visualize the radial displacement of a single segment of a patient's left ventricular myocardium.

Table 5.1 Data about hospital patients

Name	Nationality	Born	ID number	Gender	Reason for admittance
Joe Bloggs	UK	1960	123456	Male	Cardiac arrest
Zhang San	China	1977	234567	Male	Diabetes
Erika Mustermann	Germany	1986	345678	Female	Pregnancy

The files *radial1.mat* ... *radial17.mat* contain the displacement data for all 17 segments, based on the standard delineation of the myocardium into segments produced by the American Heart Association (AHA) (see Example 4.1). The files each contain two variables: `radial` represents the radial displacement measurements (in mm) and `t` represents the time (in milliseconds) of each measurement. *Both of these variables are string representations of the arrays of numeric time and displacement values.*

Write a script *m*-file that:

1. Prompts the user to enter a segment number at the command window.
2. Checks that the number entered is between 1 and 17 (the number of AHA segments). If not, the user should be continually prompted to reenter the number until a valid segment is entered.
3. Reads in the data file corresponding to the segment entered and performs any required type conversions.
4. Produces a plot of the numeric radial displacement against time. The plot should be appropriately annotated and the title should indicate the segment number.
5. Asks the user if they want to display another plot.
6. If they type 'y', repeat the above steps. Otherwise, exit.

Note that in order to read in the correct file and to display the segment number of the myocardium in the title of the plot, you will need to consider the types of the data being used and convert them if necessary. ■

5.9 THE STRUCTURE DATA TYPE

A 'structure' is a commonly used data type in a number of programming languages. It is used to wrap up information, represented by different types, into a single entity. One way of thinking about them is as a record that contains some specific fields of data that describe different aspects of an object.

For example, we might want to represent some information relating to hospital patients as given in Table 5.1.

We can build a structure to contain the details for the first patient in the list. We begin with a call to the built-in function `struct`.

```
>> patientA = struct

patientA =
struct with no fields.
```

The function returns with a message that it has constructed an empty structure. Now we can add the different fields (name, nationality, born, etc.), and assign the details for our chosen patient.

```
patientA.name = 'Joe Bloggs'
patientA.nationality = 'UK'
patientA.born = 1960
patientA.ID = 123456
patientA.gender = 'M'
patientA.reason = 'Cardiac arrest'
```

Notice the use of the dot operator to access or set each field in the structure.

We can also perform the above with a single command by passing arguments in pairs to the `struct` function. This is done in the format `field, value, ... field, value ...` to give:

```
patientA = struct('name', 'Joe Bloggs', ...
                  'nationality', 'UK', ...
                  'born', 1960, ...
                  'ID', 123456, ...
                  'gender', 'M', ...
                  'reason', 'Cardiac arrest')
```

After the assignment above, we can use the dot operator to *access* the different fields in the structure.

```
>> disp([ patientA.name ' is from ' patientA.nationality ])

Joe Bloggs is from UK
```

In the same manner, we can create structures for the other patients:

```
patientB = struct('name', 'Zhang San', ...
   'nationality', 'China', 'born', 1977, 'ID', 234567, ...
   'gender', 'M', 'reason', 'Diabetes')
patientC = struct('name', 'Erika Mustermann', ...
   'nationality', 'Germany', 'born', 1986, 'ID', 345678, ...
   'gender', 'F', 'reason', 'Pregnancy')
```

MATLAB also allows multiple structures to be collected into a single array. One way to do this is to simply concatenate them within square brackets.

```
>> patients = [patientA patientB patientC]

patients =
```

```
1x3 struct array with fields:
    name
    nationality
    born
    ID
    gender
    reason
```

MATLAB reports that it has constructed a `struct` array with three elements and lists the available fields. Note that this is only possible if the fields of all structures in the array match exactly. If they don't we need to use a cell array rather than a standard array.

We can access the fields for the elements in the array using a combination of the round brackets and the dot operator. For example, to get the names of the last two structures in the array, we can type

```
>> patients(2:3).name

ans = Zhang San
ans = Erika Mustermann
```

and MATLAB prints out the required names one after the other to the command window. Similar calls can be made for the other available fields.

■ Activity 5.8

O5.H

In Exercise 3.6 you wrote code to visualize gait tracking data and automatically identify its peaks and troughs. In this exercise you will repeat this task but using the same data supplied in two different formats. (It is not an uncommon problem in biomedical engineering to receive data in formats that are non-standard and not clearly specified!)

The files *LAnkle_tracking1.mat* and *LAnkle_tracking2.mat* contain the same data that you used in Exercise 3.6. The data types used are not specified and you will need to use MATLAB commands to work out what format the data are stored in and access the parts that you want.

You can start with your solution to Exercise 3.6 (or download the provided solution from the book's web site). Make two copies of the script *m*-file, one to work with each of the two new data files. Modify the scripts so that they perform the same tasks as specified in Exercise 3.6 but work with the new data files. To recap, the tasks are:

- Load in the data.
- Display a plot of *z*-coordinates against time.
- Automatically identify the times of all peaks and troughs of the *z*-coordinates. ■

5.10 SUMMARY

In this chapter we have looked at how a data type is used to interpret sequences of ones and zeros in the computer's memory. The same pattern of ones and zeros will be interpreted differently depending on the data type, e.g. whether it is numeric, a character, or a more sophisticated data type.

As well as looking at the fundamental data types (numeric, character and Boolean) we have looked at more advanced types such as cells and structures. We have seen that all of these can be collected into arrays and the elements can be accessed in different ways. There are yet further advanced data types that we have not considered here, such as maps and handles and more information on these can be found in the MATLAB documentation.

We have considered the range of values it is possible to represent with numeric types and how this varies. We have also looked at the precision that floating point numbers can achieve. We have introduced the special `NaN` and `Inf` identifiers to represent 'Not a Number' and 'Infinity' respectively. We have looked at ways of converting a variable between data types and how, for numeric types, MATLAB defaults to using doubles. Accepting this default behavior is generally the best advice as many functions in MATLAB expect to use doubles.

As we have seen, arrays are widely used in MATLAB and, depending on the data types of elements in the array, we have looked at the creation and access operations that can be carried out using different forms of bracket. Finally, the chapter looked at structure types and how they can be used to collect a set of details about an entity or object into a single data type.

5.11 FURTHER RESOURCES

A full list of the specific functions for testing the type of a variable can be found in the MATLAB help under *Language Fundamentals → Data Types → Data Type Identification*. Further details on conversions between data types can be found under *Language Fundamentals → Data Types → Data Type Conversion*. More ways in which structures can be created and manipulated, including conversion of data from cell arrays to arrays of structures can be found under *Language Fundamentals → Data Types → Structures*. For further description of the ASCII and Unicode encodings for characters, the Wikipedia entry is a good place to start.

EXERCISES

■ Exercise 5.1

O5.A, O5.B, O5.D, O5.E

1. Explain why a pattern of 01100011 can be viewed as the decimal number 99 or the character `'c'`.

2. Initialize your own 1×3 array of numbers. Try and choose them so that there is at least one integer that matches the ASCII code for an alphanumeric character.
3. Make an explicit conversion of the array to a character array. If some of the characters appear to be missing, why should this occur?
4. Make an explicit conversion of the array to the `logical` type. Use a MATLAB function to check that the type of the result is correct.
5. Make an *implicit* conversion of the array's contents to a character. *(Hint: We can use concatenation.)* ■

■ Exercise 5.2

O5.B, O5.D, O5.E, O5.G

1. Create a variable containing a one row character array containing `'some string'`. Do this using square brackets and individual characters. Use `class` to confirm the type of the variable.
2. Repeat the last part but create the string variable using a single set of single quotes.
3. Make a two row character array containing `'some'` and `'string'` on separate rows. What do we need to do to make sure these two words can be fitted into a two row array?
4. Make an explicit conversion of the character array from the last part to a numeric type.
5. What happens if we try to make an explicit conversion of the character array of the last part to a logical type?
6. Make an *implicit* conversion of the character array to a numeric type.
7. For each of the following strings, try and generate a corresponding numeric value using the `str2double` function. Give the reason why some of these conversions can fail and how we can test for it.

```
'2.3'
'e'
'0.7'
'X.3'
'1.3e+02'
```

■

■ Exercise 5.3

O5.D, O5.F, O5.G

1. Make a cell array called `c` that contains the following eleven strings

```
'the'
' '
'cow'
' '
'jumped'
' '
```

```
'over'
' '
'the'
' '
'moon'
```

2. Write code that accesses the *cell* at index 7 in the array `c`. Make sure you use the correct type of brackets. Use the function `class` to confirm that a cell is returned.
3. Write code that accesses the *cells* with odd indices in the array `c`, i.e. the cells at indices 1, 3, 5, etc. How many cells are returned?
4. Write code that accesses the cells with even indices in the array `c`.
5. Write code to access the *string contained* in the cell at index 5 of the array `c`.
6. Write code that accesses the last character of the string contained in the cell at index 5 in array `c`.
7. Make a new cell array `d` that contains at least three different data types. ■

■ Exercise 5.4

O5.C

1. A variable is set by the command `a = 10`. The default MATLAB behavior is to set `a` to double precision. Use the `eps` command to determine the precision of the variable.
   How would the precision change if the initial command had been `a = 1000000`? Confirm your prediction, again with the `eps` command.
2. The `intmax` function can be used to show that the largest unsigned sixteen bit integer is 65535. Show the numeric calculations that prove this is the largest possible unsigned sixteen bit integer. ■

■ Exercise 5.5

O5.F, O5.G

Here are the names of five organs: Heart, Lung, Liver, Kidney, Pancreas.

1. By padding the names with spaces where necessary, show how you can construct a 2-D `char` array to contain all the five names in strings (i.e. 1-D `char` arrays). Assign this to a variable called `a`.
2. Now show how we can put the five names into a cell array of strings. Assign this to a variable called `c`.
3. Write the code that is needed to replace Lung with Brain in the 2-D `char` array `a`.
4. Write the code that is needed to replace Lung with Brain in the cell array `c`. ■

■ Exercise 5.6

O5.B, O5.E

Consider the following definitions of numeric variables:

```
a = pi
b = uint8(250)
c = int32(100000)
d = single(17.32)
```

1. Give the full name of the type for each of the variables above.
2. An array is created by concatenating the variables above:

   ```
   X = [a b c d]
   ```

   Identify the type of the resulting array.
3. Create the following arrays by concatenating different subsets of the elements in different orders:

   ```
   X = [a d]
   X = [d a]
   X = [a d b]
   X = [a d c]
   X = [a d c b]
   X = [a d b c]
   ```

   Check the type of each result each time.
   Identify the rule that MATLAB uses for deciding on the type of the output:
   - When the array contains one or more integer types.
   - When the array contains a mixture of floating point types and no integer types.
4. A fifth variable is assigned a character: `e = 'M'`. What is the data type of the resulting array if variable `e` is included in any of the concatenations above? ■

■ Exercise 5.7

O5.F, O5.G

In Activity 2.7 you wrote code to determine the *dose escalation pattern* for a phase I clinical trial of a new cytotoxic drug. Recall that the dose escalation pattern was determined by increasing the initial dose by 67%, then by 50%, then by 40% and subsequently by 33% each time.

A common way of applying the dose escalation pattern is the "3 + 3 rule". With this approach, the trial proceeds using cohorts of 3 patients in the following way:

- If *none* of the three patients in a cohort experiences significant toxicity from the drug, another cohort of three patients will be tested at the next higher dose.
- If *one* of the three patients in a cohort experiences significant toxicity, another cohort of three patients will be treated at the *same* dose.
- The dose escalation continues until at least *two* patients among a cohort experience significant toxicity. The recommended dose for phase II of the trial is then defined as the dose just *below* this toxic dose.

The data for phase I of the cytotoxic drug trial are now available, and are contained in the file *phaseI_data.mat*. The file contains a 1 × 2 cell array. The first cell contains an array of dose levels (determined as described above) for each cohort of three patients. The second cell contains a 2-D array of `logical` values. Each row of this array represents the toxicity results of a single cohort, with `true` indicating that significant toxicity was observed in that patient, and `false` indicating no significant toxicity.

Write a MATLAB script *m*-file to determine the recommended dose for phase II of the trial from these data.

Try to make your code robust to unexpected situations, e.g. what happens if the *first* cohort has at least 2 patients who experience toxicity? You can test your code against this situation using the *phaseI_data2.mat* file provided through the book's web site. You can assume that the phase I trial continued until at least two patients in a cohort experienced significant toxicity. ■

■ Exercise 5.8

O5.D, O5.G, O5.H

In Activity 1.6 and Exercise 2.12 we introduced the concept of an Injury Severity Score (ISS). Recall that ISS quantifies trauma severity and involves each of six body regions (head, face, chest, abdomen, extremities and external) being assigned an Abbreviated Injury Scale (AIS) score between 0 and 5 depending on the severity of the trauma in that region. The three most severely injured body regions (i.e. with the highest AIS scores) have their AIS scores squared and added together to produce the ISS.

In Exercise 2.12 you wrote code to compute the ISS for a single patient from an array of numeric AIS scores. The data file *ais.mat* contains AIS scores for multiple patients in a slightly different format. Write MATLAB code to load in this file, determine the type of the data it contains, and then compute and display the ISS for all patients. ■

■ Exercise 5.9

O5.G, O5.H

Chemical elements can be described by their *name* or their *symbol*. Each element also has an *atomic number* (indicating the number of protons in its nucleus).

1. Show how a structure can be created to represent these three pieces of information for an element. Use the italics above for the field names. Give an example for an element of your choice.
2. Repeat the previous part for two more elements, and show how you can collect them into a single array of structures.
3. Show how we can access:
   - The name of the second element in the array.
   - The symbols of the first and last elements in the array. ■

FAMOUS COMPUTER PROGRAMMER: GRACE HOPPER

Grace Hopper was a pioneering American computer scientist, who due to the breadth of her achievements is sometimes referred to as "Amazing Grace". She was born in 1906 in New York City and studied mathematics and physics at undergraduate level before gaining a PhD in mathematics in 1934.

During World War II she volunteered to serve in the United States Navy. However, she was considered to be too old and not heavy enough for regular duty and so was assigned to work on the Navy's Computation Project at Harvard University, which was working to develop an early electro-mechanical computing machine. It was on this project that Grace became one of the world's first computer programmers. She is also commonly credited with popularizing the term 'debugging' for fixing programming problems (inspired by an actual moth that was removed from the computer).

After the war finished she requested to transfer to the regular Navy but was turned down due to her age. She continued to work at Harvard on the Computation Project until 1949. Grace then joined the Eckert–Mauchly Computer Corporation as a senior mathematician and joined the team developing the UNIVAC I (the second commercial computer produced in the United States). Whilst working on the UNIVAC I project she developed the world's first compiler (a program for converting a higher level language into machine readable instructions). These days almost all programming is done in such higher level languages but at the time it was a major advance and Grace had trouble convincing people of its importance. In her own words, "I had a running compiler and nobody would touch it. They told me computers could only do arithmetic." Eventually others caught on to Grace's idea, and in 1959 she was a technical consultant to the committee that defined the new language COBOL.

She always remained a Navy reservist and by the time of her retirement at the age of 79 she held the rank of rear admiral. She died in 1992, aged 85, and was interred with full military honors in Arlington National Cemetery.

> *"To me programming is more than an important practical art. It is also a gigantic undertaking in the foundations of knowledge."*
>
> **Grace Hopper**

CHAPTER 6

File Input/Output

LEARNING OBJECTIVES

At the end of this chapter you should be able to:

O6.A Use built-in MATLAB functions to save and load workspace variables

O6.B Use a range of built-in MATLAB functions to read data from external text or binary files

O6.C Use a range of built-in MATLAB functions to write data to external text or binary files

6.1 INTRODUCTION

In Section 1.6 we introduced some MATLAB commands for loading data from external files and for saving data to files. This type of operation is very common, particularly for complex programs that need to process large amounts of data. In this chapter we will look in more detail at the topic of *file input/output*.

6.2 RECAP ON BASIC INPUT/OUTPUT FUNCTIONS

First, let us recap on what we know already. Section 1.6 introduced the following simple file input/output functions:

- `save`: Save data to a MATLAB *MAT* file.
- `dlmwrite`: Write data to a delimiter-separated text file.
- `load`: Load data from a MATLAB *MAT* file or delimiter-separated text file.

Therefore, we have seen two different types of external file so far: *MAT* files and text files. *MAT* files are MATLAB specific and are used to save parts of the MATLAB workspace. When using *MAT* files the `save` and `load` functions are all we need. For the rest of Chapter 6 we will focus in more detail on functions we can use when dealing with other types of file.

MATLAB Programming for Biomedical Engineers and Scientists. DOI: 10.1016/B978-0-12-812203-7.00006-9

6.3 SIMPLE FUNCTIONS FOR DEALING WITH TEXT FILES

The `dlmwrite` command is specifically for reading numeric data from text files. Text files are files which consist of a sequence of characters, normally stored using the ASCII coding system. There are a number of other commands that are used for reading/writing numeric data from/to text files. The following are commonly used:

- `csvwrite`: Write a 1-D or 2-D array to a 'comma-separated values' text file.
- `csvread`: Read data from a comma-separated values file into a MATLAB variable.
- `dlmwrite`: Similar to `csvwrite`, but more general. Allows us to specify the delimiter used in our file, i.e. it doesn't have to be a comma.
- `dlmread`: Read delimited values from a text file.

■ Example 6.1

O6.B

For instance, consider the following example. Physiological data has been exported from an MR scanner to a file called 'scanphyslog.txt' in the format shown below.

```
-33 -53 25 -6 0 -1815 0 0 0 0000
-35 -56 24 -9 0 -1815 0 0 0 0000
-37 -59 22 -12 0 -1815 0 0 0 0000
-39 -61 20 -15 0 -1840 0 0 0 0000
-40 -61 18 -18 0 -1840 0 0 0 0000
-40 -60 17 -20 0 -1840 0 0 0 0000
-39 -58 15 -21 0 -1840 0 0 0 0000
-36 -55 15 -21 0 -1840 0 0 0 0000
-32 -50 15 -20 0 -1871 0 0 0 0000
-26 -44 17 -17 0 -1871 0 0 0 0000
-20 -37 19 -13 0 -1871 0 0 0 0000
-11 -30 23 -8 0 -1871 0 0 0 0000
-1 -22 29 -2 0 -1871 0 0 0 0000
```

The 6th column of this data file represents a signal acquired by a respiratory bellows that is sometimes used for respiratory gating of MR scans. The code example shown below will read the data from the file and then extract the bellows signal.

```
data = dlmread('scanphyslog.txt', ' ')
bellows = data(:,6)
```

Note that we specify the delimiter of the data (a space) as the second argument of `dlmread`. In fact, if we omit the delimiter, `dlmread` will attempt to use 'white space' as the delimiter, i.e. any spaces, tabs or newline characters. So, in this case, omitting the second argument would not have changed the behavior of the function. The `dlmread` function will return a 13×10 2-D array of the physiological data. The second line of code extracts the 6th column, which represents the bellows data. ■

■ Activity 6.1

The file *patient_data.txt* contains personal data for a group of patients taking part in a clinical trial. There is a row in the file for each patient, and the five columns represent: patient ID, age, height, weight and heart rate. Write a MATLAB script *m*-file that uses the `dlmread` function to read in this data. Next, write code to input a single integer value `n` from the keyboard, and display the height and weight data for the `n` oldest patients. ■

O6.B

6.4 READING FROM FILES

The functions described in Section 6.3 can be very useful but they are quite limited. Specifically,

- They only work with text files.
- They only work if the data consist of purely numeric values.

For more complex data files these functions will not be enough for our needs. In this section we will introduce some of the more flexible functions provided by MATLAB for file input/output.

Before we proceed, a key concept we should understand is the need to *open* a file before using it, and to *close* it after use. You can think of this as being similar to using a real, physical file: in this case it is obvious that the file should be opened before looking at or changing its contents. It's the same with computer files: there are specific functions that we can use to open or close external files and these must be used correctly. (Note that with the simple functions described in the previous section there was no need to make explicit calls to open or close the files – this was performed automatically for us.)

■ Example 6.2

Let us illustrate this concept with an example. The following piece of code is adapted from the MATLAB documentation for the `while` statement and it illustrates the concept of opening and closing a file as well as introducing some useful new functions.

O6.B

```
% open file
fid = fopen('test.m','r');
if (fid == -1)
    error('File cannot be opened');
end

% initialise count
```

```
count = 0;

% loop until end of file
while ~feof(fid)

    % read line from file
    tline = fgetl(fid);

    % is it blank or a comment?
    if isempty(tline) || strncmp(tline,'%',1)
        continue;
    end

    % count non-blank, non-comment lines
    count = count + 1;
end

% print result
fprintf('%d lines\n',count);

% close file
fclose(fid);
```

This code counts the number of lines in the file *test.m*, except for blank lines and comments. The `fopen` and `fclose` statements open and close a file respectively. For most MATLAB file input/output functions it is essential that we open a file before use and it is good practice to close it afterward. In addition, it is always a good idea to check that the file was opened successfully. We do this in the above code by checking the value returned by `fopen`: if it is equal to -1 this means that the file could not be opened (most likely because there was a mistake in the file name or the file was in a different location in the file system). If the file could not be opened, we use the `error` function (see Section 3.3) to report the problem and terminate the program.

The statement `feof` checks to see if the end of file has been reached and will return `true` when this happens, otherwise it returns `false`. Recall that the *tilde* operator (~) means *not* in MATLAB (see Table 2.1).

The `fgetl` function reads (gets) one line from the file. This function returns the line as an array of characters (or a *string* – see Section 5.4). Therefore, the `while` loop will continually read lines from the file so long as the end of file is not reached. When it is reached, execution passes to after the `while` loop's `end` statement.

Next, two new built-in functions are used together with an `if` statement:

- `isempty(x)`: returns whether or not the array `x` is empty.
- `strncmp(x,y,n)`: returns `true` if the strings `x` and `y` are identical up to the first `n` characters.

These two functions are used to check if the current line is either (a) empty, or (b) starts with the MATLAB comment symbol, `%`. If either of these conditions is true (recall that || means *OR* – see Table 2.1) then the `continue` statement will cause execution to pass to the beginning of the `while` loop again. If both of these tests fail, we have a line that we want to count, and 1 is added to the `count` variable. Execution then passes back to the beginning of the loop.

The `fprintf` function is used to display the number of lines to the command window. We first introduced `fprintf` in Chapter 3. It can be used as an alternative, and more flexible version, of the `disp` statement. Here it is used to display some text and a variable (`count`) to the command window (`fprintf` can also be used to write to external files, as we will see later in Section 6.5).

Finally, we make an explicit call to close the file that was opened using the `fclose` command. This tells the operating system that our code does not need to work with the file any more. ■

So, to summarize, we have learned the following new MATLAB functions for file input/output:

- `fid = fopen(filename, permission)`: Open a file called `filename` for reading/writing, giving it the file identifier `fid`. `fid` will equal −1 if the file could not be opened. The `permission` argument is a string specifying how the file will be used, e.g. 'r' for reading, 'w' for writing or 'a' for appending.
- `fclose(fid)`: Close a file that we previously opened, with file identifier `fid`.
- `line = fgetl(fid)`: Read a single line from the text file with identifier `fid`, putting the result into the string variable `line`.
- `feof(fid)`: Return a Boolean value indicating whether the end of the file with identifier `fid` has been reached (`true`) or not reached (`false`).

Now let us consider another situation. Example 6.2 illustrated how entire lines could be read from a text file as strings (1-D arrays of characters). However, it is often the case that a line of data contains a number of individual values which should be stored separately for further processing. In such cases a more flexible file input function is needed.

■ Example 6.3

In this example we would like to read some data about patients' blood pressures from a text file. The file *sys_dia_bp.txt* contains the following data: *O6.B*

```
BP data
51 85
88 141
67 95
```

```
77 111
68 115
99 171
80 121
```

First, there is a header line which just contains the text "BP data". After this, in each row of the file, the value in the 1st column represents the patient's systolic blood pressure and the value in the 2nd column represents the diastolic blood pressure of the same patient. The following piece of MATLAB code can be used to read all of the data from the file into a single 2-D array called `data`.

```
% open file
fid = fopen('sys_dia_bp.txt', 'r');
if (fid == -1)
    error('File cannot be opened');
end

% skip header line
line = fgetl(fid);

% read systolic and diastolic blood pressure
data = fscanf(fid, '%d %d', [2 inf])

% close file
fclose(fid);
```

Note that, as before, we open the file before reading any data, and close it after we have finished reading. We use the `fgetl` function that we saw in Example 6.2 to read (and discard) the text header line. Next, in a single line of code, the `fscanf` function is used to read all rows of data from the file. Three arguments are provided to `fscanf`:

- The identifier of the file containing the data: `fid`.
- A *format string* that specifies the format of the data fields to read: `%d` means a signed decimal (base 10) number. `'%d %d'` represents two numbers, so `fscanf` will read numbers from the file in pairs (i.e. a row at a time).
- The dimensions of the output array returned by `fscanf`: `[2 inf]` means that the data will be placed in an array with two rows (representing the pairs of numbers being read in) and an unlimited number of columns, i.e. the function will read pairs of decimal numbers until it reaches the end of the file.

In this example, seven pairs of decimal numbers will be read in, as there are seven lines of numeric data in the input file. Therefore, the output variable `data` will be a 2×7 array. ■

Table 6.1 Field specifiers for use in the format strings of file input/output functions (e.g. `fscanf`)

Specifier	Field type
`%d`	Signed decimal integer
`%u`	Unsigned decimal integer
`%f`	Floating point number
`%c`	Character
`%s`	String (i.e. a sequence of characters)

The *format string* we saw in the above example was first introduced in Section 3.3. As well as reading decimal numbers, format strings enable `fscanf` to be used to read in a range of other data types. A summary of some of the more common *field specifiers* that can be used in the format string is provided in Table 6.1. Note, however, that `fscanf` cannot be used to read *mixed* data types from a single file. For example, it is not possible to read a string followed by a decimal integer using `fscanf`. The reason for this is that `fscanf` always returns an array with elements of one basic data type, i.e. one that cannot contain elements with different data types. It is possible to read a file containing differing *numeric* data types (e.g. a signed integer and a floating point) but in this case `fscanf` will perform an automatic conversion of one of the types to ensure that the resulting array contains only a single type (e.g. it will convert the signed integer to floating point numbers).

It was in fact possible to perform the work of Example 6.3 using the `dlmread` command, as the following code illustrates.

```
data = dlmread('sys_dia_bp.txt', ' ', 1, 0)
```

The final two numerical arguments of `dlmread` indicate that the data should be read starting from row 1, column 0 (i.e. it should skip the first row). However, `fscanf` is more flexible than `dlmread` and can be used to read a wider range of data formats, as the following example illustrates.

■ Example 6.4

Suppose now that the file *sys_dia_bp2.txt* contains the following data: O6.B

```
BP data
Sys 51 Dia 85
Sys 88 Dia 141
Sys 67 Dia 95
Sys 77 Dia 111
Sys 68 Dia 115
Sys 99 Dia 171
Sys 80 Dia 121
```

This is the same numerical data that we saw in Example 6.3 but with extra text between the numerical values. In biomedical engineering it is common to have to deal with data exported from a range of devices with different data formats, so this example is quite realistic. It would not be straightforward to use `dlmread` to read this data because each row of data has mixed data types, i.e. strings and numbers. However, with `fscanf` the solution is quite simple.

```
% open file
fid = fopen('sys_dia_bp2.txt');
if (fid == -1)
    error('File cannot be opened');
end

% skip header line
line = fgetl(fid);

% read systolic and diastolic blood pressure
data = fscanf(fid, '%*s %d %*s %d', [2 inf])

% close file
fclose(fid)
```

All we have done here is to modify the format string of the `fscanf` function: we have added in two extra string fields before each decimal integer field (these correspond to the `Sys` and `Dia` strings in the data file). Note the * symbol between the `%` and `s`: this causes MATLAB to read the specified field but to exclude it from the output array. This is necessary because, although it can read files containing mixed data types, `fscanf` can only return an array of values of a single type. ■

■ Activity 6.2

O6.B

This exercise builds on Activity 6.1, in which you wrote MATLAB code to read in patient data from the text file *patient_data.txt*. On the book's web site you will also be able to access a second patient data file called *patient_data2.txt*. This also contains patient data (patient ID, age, height, weight, heart rate) but for some different patients in the trial. Modify the script *m*-file you wrote in Activity 6.1 to also read in this data and combine it with the data from the first file. There were five pieces of data for each of nine patients in the first file and the same data for an additional five patients in this second file, so your final data should be a 14×5 array. Notice that this second file has no line break characters so you will not be able to use `dlmread`, because this uses line breaks to separate the rows of data. ■

Sometimes, however, it is desirable to read in data of mixed types and store it all in a single variable. To do this, we need an alternative file input function: `textscan`. The following example illustrates its use.

■ Example 6.5

Suppose now that the file *sys_dia_bp3.txt* contains the patients' names as well as their blood pressure data, and that we wish to read in and store all of the data, i.e. names and blood pressures:

O6.B

```
Joe Bloggs 51 85
Josephine Bloggs 88 141
Zhang San 67 95
Anders Andersen 77 111
Erika Mustermann 68 115
Ola Nordmann 99 171
Jan Kowalski 80 121
```

It is not possible to read and store all of this data using `fscanf` because the resulting array could not contain both string and numeric data. However, `textscan` returns a *cell array* rather than a normal array (see Section 5.7), and cell arrays *can* contain mixed data types. Therefore, it is possible to read this data using `textscan`. Examine the code shown below.

```
% open file
fid = fopen('sys_dia_bp3.txt', 'r');
if (fid == -1)
    error('File cannot be opened');
end

% read systolic and diastolic blood pressure
data = textscan(fid, '%s %s %d %d');

% close file
fclose(fid);
```

Note that `textscan` has a similar format string to `fscanf`. However, in this case it is permitted to mix up string data fields (`%s`) with decimal number fields (`%d`) and all fields will be included in the returned variable (`data`). The type of the variable `data` returned by `textscan` will now be a 1×4 *cell* array.

The first and second elements of `data` are arrays that contain the first names and second names of the patients respectively. In fact, since the lengths of the strings vary these first two elements are themselves cell arrays. The third and fourth elements of `data` are normal arrays of integers representing the systolic and diastolic blood pressures of the patients. Try entering this code into a MATLAB script *m*-file, running it and inspecting the `data` variable. ■

The `textscan` function is quite flexible and can also be used to read in data of mixed types and with multiple delimiters.

■ Example 6.6

O6.B

For example, let's change the format of the data, in a new file called *sys_dia_bp4.txt*:

```
Joe Bloggs;51;85
Josephine Bloggs;88;141
Zhang San;67;95
Anders Andersen;77;111
Erika Mustermann;68;115
Ola Nordmann;99;171
Jan Kowalski;80;121
```

If we were to use the code shown in Example 6.5 to read in this data it would read it incorrectly because the default delimiter for `textscan` is white space, i.e. spaces, tabs or newline characters. As a result the 2nd, 3rd and 4th data fields (i.e. the last name, systolic blood pressure and diastolic blood pressure) would be read as a single string. The following code shows how we can change the delimiter(s) for `textscan`.

```
% open file
fid = fopen('sys_dia_bp4.txt', 'r');
if (fid == -1)
    error('File cannot be opened');
end

% read systolic and diastolic blood pressure
data = textscan(fid, '%s %s %d %d', 'delimiter', ' ;');

% close file
fclose(fid);
```

Take care to note that the string used to specify the delimiter contains multiple characters (a space and a semi-colon). This appears after the `'delimiter'` argument. In this case, either of these two characters will be treated as a delimiter, and this allows the data fields to be read in correctly as before. ■

■ Activity 6.3

O6.B

The file *names.txt* contains the first and last names of a group of patients. Write a MATLAB script *m*-file to read in this data using the `textscan` function and display the full names of all patients with the last name "Bloggs".

(Hint: Look at the MATLAB documentation for the `strcmp` command.) ■

So far, all our examples have read data from, or written data to, text files. In computer programming, it is common to distinguish between text files and *binary* files. In fact, strictly speaking, all computer files are binary files since the data are always stored as a series of 1s and 0s (known as binary digits, or *bits*).

However, it is common to refer to files as text files if the sequence of bits can be interpreted as character data using a coding system such as ASCII. Any other file is referred to as a binary file. Binary files tend to make more efficient use of disk space than text files, but can occasionally be harder to read. Regardless of the advantages and disadvantages of text and binary files, it is a fact that both types of file are in common use so we need to know how to read from and write to both.

We have already seen how the `load` and `save` commands can be used with *MAT* files, which are binary files. We will now consider other functions that can be used with binary files and that allow us more control over how the data are interpreted. The main function to know about for reading data from binary files is `fread`. The use of `fread` is illustrated in the following example.

■ Example 6.7

Suppose that the file *heart_rates.bin* is a binary file containing the following heart rate data stored as a series of 16-bit integers: `[65 73 59 101 77 ... 90 68 92]`. How can we write code to read these values into a MATLAB array? Consider the program listing shown below.

O6.B

```
% open file
fid = fopen('heart_rates.bin', 'r');
if (fid == -1)
    error('File cannot be opened');
end

% read data from binary file as 16-bit integers
hr = fread(fid, inf, 'int16')

% close file
fclose(fid);
```

In this example, the `fread` function takes three arguments:

- The identifier of the binary file from which to read the data: `fid`.
- The number of data values to read: using `inf` means read until the end of the file.
- The data type of the elements to be read: `int16` specifies 16-bit integers (see Section 1.5).

Note that if we didn't specify `int16` as the data type `fread` would assume that each byte (8 bits) represented a number and so the data would not be read correctly. Generally, when reading data from binary files it is important to know and to specify the type of the data we are reading since different data types use different numbers of bytes on the computer's hard disk. ■

■ Activity 6.4

The file *cholesterol.bin* is a binary file. The first data element in the file is

O6.B

an 8-bit integer specifying how many records follow subsequently. Each subsequent record consists of an 8-bit integer representing a patient's age followed by a floating-point `double` representing the same patient's total cholesterol level (in mmol/L). Write a MATLAB script *m*-file to read the age/cholesterol data into two array variables. ■

6.5 WRITING TO FILES

Now we will introduce some MATLAB functions that can be used for writing data to binary or text files. For binary files, the main function for writing data is `fwrite`, the use of which is illustrated in the following example.

■ Example 6.8

O6.C

The code shown below can be used to generate the binary file that was read in using `fread` in Example 6.7.

```
% define data array
hr = [65 73 59 101 77 90 68 92];

% open file
fid = fopen('heart_rates.bin', 'w');
if (fid == -1)
    error('File cannot be opened');
end

% write data to binary file as 16-bit integers
fwrite(fid, hr, 'int16');

% close file
fclose(fid);
```

In this example, we provide three arguments to `fwrite`:

- A file identifier: `fid`.
- An array containing data to be written to the file: `hr`.
- The data type to use when writing: `int16`.

Note that there is no argument for specifying how much of the array should be written – the entire array will be written. If we only want to write part(s) of the array we need to modify the value of the second argument ourselves, or use multiple `fwrite` commands. Note also that using a single `fwrite` command, only values of a single data type can be written. If we want to write mixed data to a binary file we can use multiple successive `fwrite` commands before closing the file. ■

The main function for writing data to text files is `fprintf`, which we saw earlier in Example 6.2. The `fprintf` function is actually very powerful and flexible, and can be used to write data of a range of different data types, as well as specify the formatting of that data. Consider the following example.

■ Example 6.9

The code shown below can be used to generate the mixed data file that was read in using `textscan` in Example 6.5.

O6.C

```
% define cell array of names
names = {'Joe Bloggs','Josephine Bloggs', ...
         'Zhang San','Anders Andersen', ...
         'Erika Mustermann','Ola Nordmann', ...
         'Jan Kowalski'};

% define array of blood pressure values
bps = [51 85
       88 141
       67 95
       77 111
       68 115
       99 171
       80 121];

% open file
fid = fopen('sys_dia_bp3.txt', 'w');
if (fid == -1)
    error('File cannot be opened');
end

% write data to file
for x=1:length(names)
    fprintf(fid, '%s %d %d \n', names{x}, bps(x,1), bps(x,2));
end

% close file
fclose(fid);
```

Here we use a cell array called `names` to store the names of the patients. Each element of `names` is a normal 1-D array of characters. Because the names have different lengths we can't store them using a normal 2-D character array. The blood pressure values are stored using a normal 2-D array of numbers. After opening the file for writing, we use a `for` loop to iterate through all patients (determined by finding the length of the `names` cell array). In each iteration, the `fprintf` function is used to construct and write a line to the file. This is formed from a string (patient's name), followed by two decimal values (the systolic and diastolic blood pressures).

We supply five arguments to `fprintf` in this example:

- The file identifier: `fid` – if this is omitted the line is displayed in the command window as we saw in Example 6.2.
- The format string, in this case containing three fields (a string and two decimals), and ending with a new-line character (`'\n'`).

- The remaining three arguments contain the data for each field in the format string – so `names{x}` is used for the string field, `bps(x,1)` for the first decimal and `bps(x,2)` for the second decimal.

Finally, after the loop has finished executing, the output file is closed. ■

When specifying a format string for `fprintf`, we can use the same field specifiers that we use with `fscanf` (see Table 6.1). In addition, we can provide extra information to alter the display format of the fields by adding extra characters in between the `%` and the data type identifier. For example, we can specify the width and precision of numeric fields, e.g.

```
fprintf(fid, '%10.3f \n', x);
```

This code will display the value of the variable `x` as a floating point number using a field width of 10 characters and a precision of 3 (i.e. 3 numbers after the decimal point). For example, if the variable `x` contained the square root of two, the line produced by `fprintf` above would consist of the string: '   1.414' which contains five spaces before the digit 1.

In addition, a number of flags can be added before the field width/precision values. Adding the – character will cause the displayed number to be left-justified (the default is right-justified). Adding the + character will cause a plus sign to be displayed for positive values (by default only a minus sign is added for negative values). Adding a space or `0` will cause the number to be left-padded with spaces or zeroes up to the field width. For example, the following statement,

```
fprintf('%010.3f \n', x);
```

is the same as that shown above except that the number will be padded to the left with zeroes to make the field width 10 characters ('000001.414').

■ Activity 6.5

O6.B, O6.C

The file *bp data.txt* contains systolic blood pressure data. There are four fields for each patient: patient ID, first name, last name and systolic blood pressure. Write a MATLAB script *m*-file to read in this data, identify any 'at risk' patients (those who have a blood pressure greater than 140) and write only the data for the 'at risk' patients to a new file, called *bp_data_at_risk.txt*. ■

6.6 SUMMARY

MATLAB provides a range of file input and output functions for interacting with external files. Each function can be useful in different situations, depending on the type(s) of the data, the format of the file, and whether we are dealing

Table 6.2 Summary of file input/output functions

Function	File type	Operation	Open/close file?
`load`	Binary[a] or text	Read MATLAB variables	N
`save`	Binary[a]	Write MATLAB variables	N
`csvread`	Text	Read comma separated value numeric data from a file into a 1-D or 2-D array	N
`dlmread`	Text	Read delimiter separated value numeric data from a file into a 1-D or 2-D array	N
`csvwrite`	Text	Write 1-D/2-D array of numeric data to a file: comma separated value format	N
`dlmwrite`	Text	Write 1-D/2-D array of numeric data to a file: general delimiter	N
`fread`	Binary	Read data of a single type from a file into a 1-D array	Y
`fscanf`	Text	Read data of a single type from a file into a 1-D/2-D array	Y
`textscan`	Text	Read a file containing mixed data types into 1-D/2-D cell array	Y
`fgetl`	Text	Read a line of data into a string (1-D array of characters)	Y
`feof`	Text	Check for the end-of-file condition	Y
`fwrite`	Binary	Write an array of data of a single type to a file	Y
`fprintf`	Text	Write array(s) of data of mixed data types to a file	Y

[a] MAT *file.*

with a binary or a text file. Table 6.2 summarizes the key characteristics of the functions we have discussed in this chapter.

6.7 FURTHER RESOURCES

- MATLAB documentation on text file functions: http://www.mathworks.co.uk/help/matlab/text-files.html.
- MATLAB documentation on general file input/output functions: http://www.mathworks.co.uk/help/matlab/low-level-file-i-o.html.

EXERCISES

■ Exercise 6.1

Write a MATLAB script *m*-file to create the following variables in your MATLAB workspace, and initialize them as indicated: O6.A

- a character variable called `x` with the value 'a'.
- an array variable called `q` containing 4 floating point numbers: 1.23, 2.34, 3.45 and 4.56.

- a Boolean variable called `flag` with the value `true`.

Now save only the array variable to a *MAT* file, clear the workspace, and load the array in again from the *MAT* file. ■

■ Exercise 6.2

O6.B, O6.C

Write the same 1-D array of floating point numbers that you defined in Exercise 6.1 to a text file using the `csvwrite` function. Then clear the workspace and read the array in again using `csvread`. ■

■ Exercise 6.3

O6.B, O6.C

Again, working with the array from Exercise 6.1, this time write it to a text file using the `dlmwrite` function. Use the colon character as the delimiter and write the values with a precision of 3 decimal places. Then clear the workspace and read the array in again using `dlmread`.

(Hint: Look at the MATLAB documentation for `dlmwrite` *to find out how to specify the precision of the data.)* ■

■ Exercise 6.4

O6.B

The equation for computing a person's body mass index (BMI) is:

$$\text{BMI} = \frac{\text{mass}}{\text{height}^2}$$

where the mass is in kilograms and the height is in meters. A BMI of less than 18.5 is classified as underweight, between 18.5 and 25 is normal, more than 25 and less than 30 is overweight, and 30 or over is obese. You are provided with two text files: *bmi_data1.txt* and *bmi_data2.txt*. These both contain data about patients' height and weight (preceded by an integer patient ID), but in different formats. Write a MATLAB script *m*-file to read in both sets of data, combine them, and then report the patient ID and BMI classification of all patients who do not have a normal BMI. ■

■ Exercise 6.5

O6.B

When MR scans are performed a log file is typically produced containing supplementary information on how the scan was acquired. An example of such a file (*logcurrent.log*) is available to you. One piece of information that this file contains is the date and time at which the scan started. This information is contained in the 2nd and 3rd words of a line in the file that contains the string "Scan starts". Write a MATLAB script *m*-file to display the start date and time for the scan which produced the *logcurrent.log* file.

(Hint: Look at the MATLAB documentation for the `strfind` *and* `strsplit` *functions.)* ■

■ Exercise 6.6

O6.B

Whenever a patient is scanned in an MR scanner, as well as the log file mentioned in Exercise 6.5, the image data are saved in files that can then be exported for subsequent processing. One format for this image data, which is used on Philips MR scanners, is the PAR/REC format. With PAR/REC, the image data themselves are stored in the REC file whilst the PAR file is a text file that contains scan details to indicate how the REC file data should be interpreted. An example of a real PAR file (*patient_scan.par*) is available to you through the book's web site. Write a MATLAB script *m*-file to extract from the PAR file and display the following information: the version of the scanner software (which is included at the end of the 8th line), the patient name and the scan resolution. ■

■ Exercise 6.7

O6.B, O6.C

Measuring blood sugar levels is an important part of diabetes diagnosis and management. Ten patients had their post prandial blood sugar level measured and the results were 5.86, 8.71, 4.83, 7.05, 8.25, 7.87, 7.14, 6.83, 6.38 and 5.77 mmol/L. Write a MATLAB script *m*-file to write these data to a binary file using an appropriate data type and precision. Your program should then clear the MATLAB workspace and read in the data again. ■

■ Exercise 6.8

O6.B, O6.C

Write a MATLAB script *m*-file to read in a text file and write out a new text file that is identical to the input file except that all upper case letters have been converted to lower case. You can test your code using the file *text.txt*. ■

■ Exercise 6.9

O6.B, O6.C

Write a MATLAB script *m*-file to read in another *m*-file, ignoring all lines that are blank or contain comments. It should write the remaining statements to a new *m*-file. Note that a comment is any line in which the first non-white space character is a `%`.

(Hint: The MATLAB command `strtrim` can be used to remove leading and trailing white space from a string.) ■

FAMOUS COMPUTER PROGRAMMER: EDSGER DIJKSTRA

Edsger Dijkstra was a Dutch computer scientist, who was born in 1930 in Rotterdam. Originally, he wanted to study law, and had the ambition to represent the Netherlands at the United Nations. However, his family persuaded him to turn his attentions to science. He then planned to become a theoretical physicist, and went to study mathematics and physics at the University of Leiden.

In 1951, a chance event changed Edsger Dijkstra's life. He saw an advertisement for a 3-week course on computer programming at the University of Cambridge in the UK, and he decided to attend. Due to the scarcity of such skills at the time, this resulted in Dijkstra getting a part-time job as a programmer in the Computation Department of the Mathematical Center in Amsterdam. Subsequently, he reflected that "in '55 after three years of programming, while I was still a student, I concluded that the intellectual challenge of programming was greater than the intellectual challenge of theoretical physics, and as a result I chose programming". From that moment on he dedicated his life to computer science.

Dijkstra married in 1957, and was required to state his profession on the marriage certificate. He wrote that he was a Computer Programmer, but this was not accepted because the authorities claimed that there was no such profession. He had to change it to Theoretical Physicist.

In 1962 Dijkstra became a Professor of Mathematics at Eindhoven University, because it had no Computer Science department at the time. Whilst there, he built a team of computer scientists, and performed some pioneering work on computers and computer programming. He is considered to be one of the very earliest computer programmers and has been one of the most influential figures in computer science since its inception.

Despite having been responsible for much of the technology underlying computer programming, Dijkstra resisted the personal use of computers for many decades. He would produce most of his articles by hand, and when teaching he would use chalk and a blackboard rather than overhead slides.

Edsger Dijkstra died in 2002 at the age of 72.

> *"How do we convince people that in programming simplicity and clarity – in short: what mathematicians call 'elegance' – are not a dispensable luxury, but a crucial matter that decides between success and failure?"*
>
> **Edsger Dijkstra**

CHAPTER 7

Program Design

LEARNING OBJECTIVES

At the end of this chapter you should be able to:

O7.A Use structure charts and pseudocode to perform a top-down stepwise refinement design of a complex problem
O7.B Convert a top-down design into working MATLAB code
O7.C Use an incremental testing approach to program development and verification, making appropriate use of test stubs
O7.D Maximize code reuse when designing and developing software

7.1 INTRODUCTION

So far we have learned how to make use of a range of MATLAB programming features to develop increasingly sophisticated programs. In Chapter 3 we looked at how to define our own functions for performing specific tasks. However, as we start to address more and more complex problems it is unlikely that they will be easily solved by writing one big function. Rather, an efficient and elegant implementation is likely to consist of a number of functions that interact by making calls to each other and passing data as arguments. *Program design* refers to the process of deciding which functions to write and how they should interact.

When tackling a larger programming problem, it is generally not a good idea to start coding straight away: time invested in producing and documenting a high quality structured design will produce significant benefits in the long run, primarily by reducing time spent writing code and maintaining it. This chapter will introduce some tools that will be useful when designing programs to solve larger and more complex problems.

We also return to the important topic of program testing. Testing, or verifying, that a program meets its requirements is an essential part of the development process. In Chapter 4 we saw how an incremental approach to writing programs, with testing of the incomplete program after each stage, can actually

MATLAB Programming for Biomedical Engineers and Scientists. DOI: 10.1016/B978-0-12-812203-7.00007-0

speed up and improve the development process. In this chapter we will discuss how this philosophy of incremental development interacts with a structured design process.

In computer science, it is common to identify two distinct approaches to program design: *top-down* design and *bottom-up* design. In this chapter we will discuss the meanings of these two concepts. However, in procedural programming (see Section 1.1.1), which is the subject of this book, top-down design has been the traditional approach, so we will focus mainly on this, illustrating it with a simple case study.

7.2 TOP-DOWN DESIGN

With a top-down design approach, the emphasis is on planning and having a complete understanding of the system before writing code. Top-down design is also often referred to as *stepwise refinement*. The basic idea is to take a hierarchical approach to identifying program 'modules'. By a 'module' we simply mean a part of the overall program that performs a specific task. For example, in MATLAB, we can consider a function to be a 'module'.

In top-down design, we start off by breaking down a complex problem into a number of simpler sub-problems. Each sub-problem will be addressed by a separate program module, which will have its requirements separately specified, including how it should interact with the *main* part of the program (i.e. the top level of the hierarchy). The interaction is typically specified in terms of what data are passed to a given module from the main module, and what data are passed back from the module to the main module.

Next, each identified module can be broken down further into sub-modules, which will also have their behavior specified, including what data are passed between the new sub-modules and their parent modules. This process continues until the sub-modules are simple enough to be easy to implement (i.e. to write as code).

The result of this *stepwise refinement* is a hierarchy of modules and sub-modules that represents the structure of the program to be implemented. This, together with specifications for the behavior of each module, represents the program design. Normally, no coding will begin until a sufficient level of detail has been reached in the design of at least some part of the system. Often, the complete structured design will be produced before any coding takes place.

We will illustrate this process with a case study.

■ Example 7.1

O7.A, O7.B, O7.C, O7.D

The file *patient_data.txt* contains data for a group of patients registered at a GP surgery (patient ID, name, date of birth). A second file, *patient_examinations.txt*, contains data about physical examinations that patients at the surgery have undergone (patient ID, age, height, weight, systolic/diastolic blood pressure, total cholesterol level). There may be multiple examinations for each patient, or none.

The GP has decided to implement a program in which 'high risk' patients are invited for further tests. The initial definition for 'high risk' patients is those who have an age that was greater than or equal to 75 at their most recent examination.

We are required to design and implement a program to read in the data from these files and display the patient IDs and names of all high risk patients. If there are no high risk patients the program should terminate with an appropriate message. ■

We will start by breaking the problem down into simpler sub-problems. In software design, this is known as *factoring*, and the first time we apply it to the overall problem it is known as *first level factoring*. We will illustrate our first level factoring with a commonly used graphical technique known as *structure charts*.

Step 1 – First Level Factoring

Fig. 7.1a shows the structure chart after first level factoring has been performed. The main program (which we have called *Report Risk*) is at the top level of the hierarchy. Below this, we have broken down our initial problem into three sub-problems. We are saying that the requirements of our program can be met by executing three modules one after the other: first we read all personal and examination data from the files (*Read Data*), then we perform the risk screening (*Select High Risk Patients*), and finally we display the details of the selected patients (*Report High Risk Patients*). The arrows that link the three sub-modules to the parent (main) module are annotated with the data that needs to be passed between them. Empty circles with an arrow indicate data flowing in the direction of the arrow. For example, the *Read Data* module should pass the personal and examination data that was read back to the *Report Risk* program. These data are then passed to the *Select High Risk Patients* module for screening, which returns a list of the selected high risk patients. These data are passed to the *Report High Risk Patients* module.

Currently, none of these three new modules are simple enough to be implemented easily, so we proceed by further breaking down one of them (*Read Data*). This is known as *further factoring*.

Step 2 – Further Factoring

Fig. 7.1b shows the additional structure chart for the further factoring of *Read Data*. Now, we are saying that the problem of reading in all data can be solved by first reading in the personal data and then reading in the examination data.

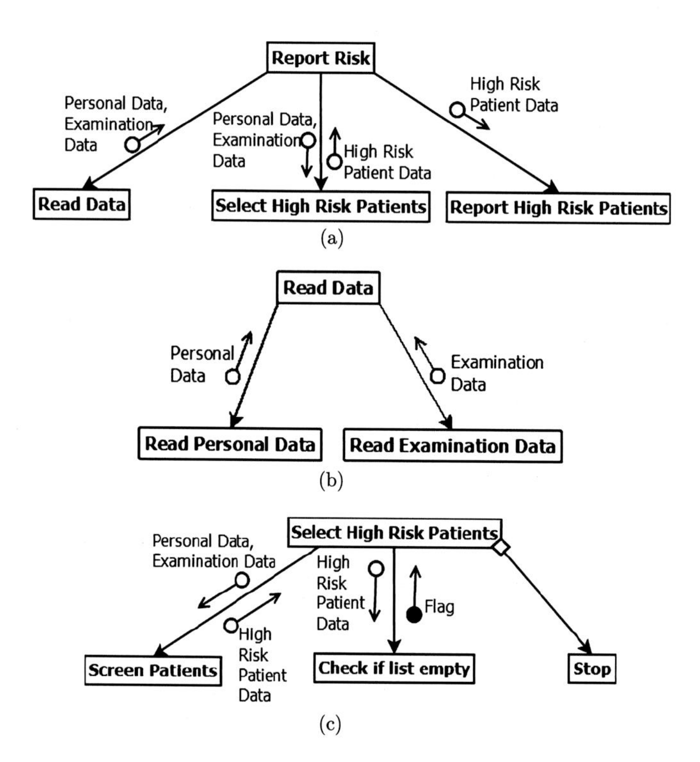

FIGURE 7.1 Structure charts representing the design of the risk screening program: (a) Step 1 – first level factoring; (b) Step 2 – further factoring of *Read Data*; (c) Step 3 – further factoring of *Select High Risk Patients*. Empty circles represent data, filled circles represent control. The diamond in (c) represents conditional execution.

Note the data flowing between the parent module (*Read Data*) and its two sub-modules.

Step 3 – Further Factoring

Now we proceed with further factoring of the *Select High Risk Patients* module identified in **Step 1**. Fig. 7.1c shows the structure chart that represents the operation of this module. This specifies how the problem of selecting the high risk patients can be solved: first, the screening operation is performed, then we check if the list of patients is empty or not, and finally the program stops if the list was empty. This time there is some new notation in the structure chart. In addition to the empty circles with arrows representing data, we also have a filled circle. A filled circle with an arrow in a structure chart indicates *control* information. Control information is normally a binary value that has some effect on future program execution. For example, in Fig. 7.1c the control information called `Flag` is passed back from the *Check if list empty* sub-module: this will be true if there are no high risk patients and false otherwise. This value

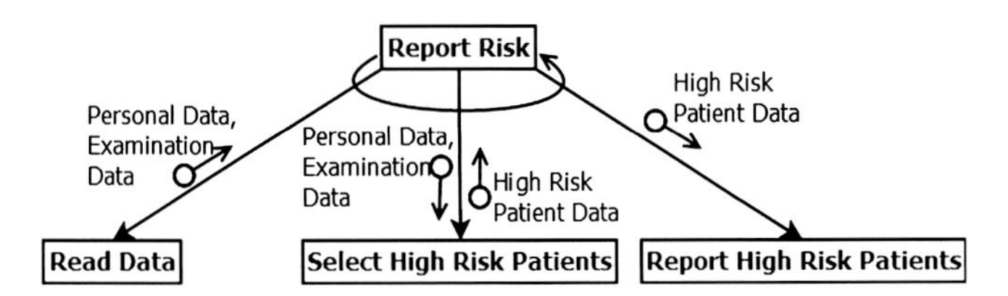

FIGURE 7.2 Structure chart showing iteration of sub-modules.

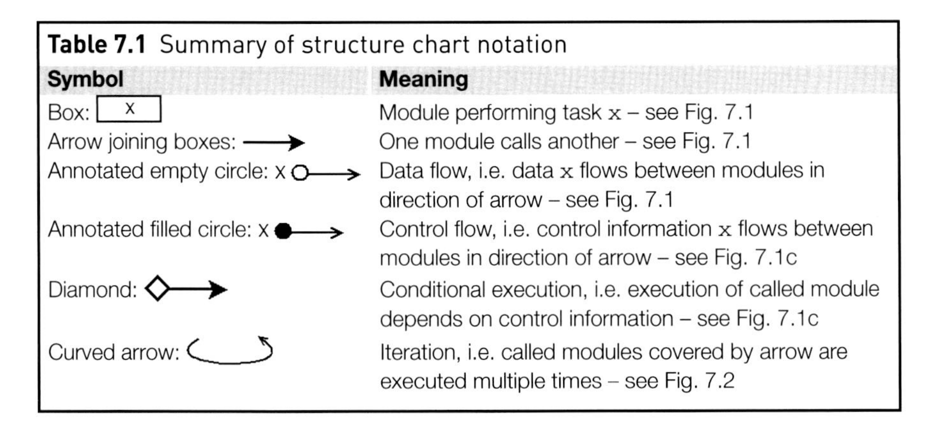

Table 7.1 Summary of structure chart notation

Symbol	Meaning
Box: [x]	Module performing task x – see Fig. 7.1
Arrow joining boxes: ⟶	One module calls another – see Fig. 7.1
Annotated empty circle: x ○⟶	Data flow, i.e. data x flows between modules in direction of arrow – see Fig. 7.1
Annotated filled circle: x ●⟶	Control flow, i.e. control information x flows between modules in direction of arrow – see Fig. 7.1c
Diamond: ◇⟶	Conditional execution, i.e. execution of called module depends on control information – see Fig. 7.1c
Curved arrow: ↻	Iteration, i.e. called modules covered by arrow are executed multiple times – see Fig. 7.2

is then used to decide whether or not to execute the next sub-module, *Stop*. The *conditional execution* of the *Stop* sub-module is indicated by the diamond symbol where its line joins the *Select High Risk Patients* module.

As an aside, as well as conditional execution (which corresponds to conditional statements in procedural languages such as MATLAB, such as `if` statements – see Chapter 2), structure charts can also represent *iteration*. For instance, suppose that we wanted our program to screen data from multiple GP surgeries, i.e. the *Read Data, Select High Risk Patients* and *Report High Risk Patients* modules together should be iteratively executed. Fig. 7.2 shows how the structure chart shown in Fig. 7.1a should be modified to indicate this iteration. The curved arrow covering the three modules indicates that they should be executed multiple times.

Table 7.1 provides a summary of the notation we have introduced for structure charts.

Step 4 – Write Pseudocode

To return to our case study, we now proceed by examining which modules might need further elaboration. Looking at Fig. 7.1b (i.e. the *Read Data* module) we might conclude that this structure chart has enough detail already to proceed to implementation. However, there are two other modules that do require further analysis: *Report High Risk Patients* from the first-level factoring, and the new sub-module from **Step 3** (*Screen Patients*). Although it would be possible to continue to use structure charts to further break down these problems, we will now introduce a new way of representing program structure: *pseudocode*. Pseudocode is simply an informal description of the steps involved in executing a computer program, often written in something similar to plain English. Pseudocode is commonly used when we get down to the lower-level details of how modules work internally, rather than the interactions of higher-level operations commonly represented by structure charts.

For example, the pseudocode below shows more detail on how the *Screen Patients* module works (which was identified in the structure chart shown in Fig. 7.1c).

PSEUDOCODE – Screen Patients:

```
For each PatientID in Personal Data
  Look up all records in Examination Data with Patient ID
  HighRisk = false
  For each record
    If Age >= 75
      HighRisk = true
  If HighRisk = true
    Add PatientID, Name to High Risk Patients List
Return High Risk Patients List
```

This pseudocode indicates the steps involved in screening for high risk patients. We first iterate over all patients. For each patient, we look up all corresponding examination records (i.e. those with the same Patient ID). Next, we initialise a `HighRisk` variable to false. Then, for each of the corresponding examination records, we check the age of the patient at the time of the examination, and if it is greater than or equal to 75 we set `HighRisk` to true. This way, if the age at *any* examination was greater than or equal to 75 then `HighRisk` will be true, otherwise it will be false. If `HighRisk` is true we add the patient details to the high risk patients list. Finally, we return the list as the output of the module.

Note that this pseudocode is starting to look more like computer code, with variables being defined and simple operations to act on them. However, we haven't yet written any code, so our design could be implemented in any language of our choice, including MATLAB.

Pseudocode for the remaining module of our design (*Report High Risk Patients*) is shown below.

PSEUDOCODE – Report High Risk Patients:

```
For each record in High Risk Patients List
  Display PatientID, Name
```

In summary, a typical top-down design process would involve first level factoring (e.g. Fig. 7.1a), followed by further factoring (e.g. Figs. 7.1b–c) as needed until a certain level of detail is reached. Then some of the bottom level modules in the structure charts would be expanded by writing pseudocode. The choice of when to switch from structure charts to pseudocode is subjective and a matter of personal preference: you may prefer to use entirely structure charts, or entirely pseudocode, but it is common to use a mixture of the two as we have in this example. Either way, once the design is complete and documented, coding can begin.

7.2.1 Incremental Development and Test Stubs

We now have a design that can be used as the starting point for producing an implementation. Regardless of the programming language used, we need a general approach for producing our implementation. In Section 4.2 we introduced the concept of *incremental development*, in which software is developed in small pieces, with frequent testing of the developing program. In Example 4.1 a program for cardiac dyssynchrony index calculation was developed incrementally, and the entire program was tested after each new piece of code was added. However, this was for a relatively simple example in which each part of the overall program consisted of just a few lines of code. Now that we are tackling larger, more complex problems, it is more likely that our operations will be functions that perform quite sophisticated tasks. So this begs the question: how can we perform incremental development and testing when implementing a top-down design of a larger program?

The answer to this question lies in the concept of a *test stub*. A test stub is a simple function or line of code that acts as a temporary replacement for more complex, yet-to-be-implemented code. For instance, in our case study we might use test stubs for the *Select High Risk Patients* and *Report High Risk Patients* modules whilst we are developing and testing the code for the *Read Data* module. The test stub for *Select High Risk Patients* could, for example, always return a list containing a single patient, which could be 'hard-coded', i.e. just directly assigned to the appropriate variable. The test stub for *Report High Risk Patients* could always display a simple message. The code listing shown below illustrates such a program under development.

ReportRisk.m:

```
% GP risk screening program

% Read Data (under development in another file)
[personalData, examData] = ReadData();

% Select High Risk Patients (test stub)
highRisk = '123 Joe Bloggs';

% Report High Risk Patients (test stub)
fprintf('High Risk:\n');
```

Once the *ReadData* module has been developed and tested, the test stub for *Select High Risk Patients* could be replaced with an implementation and tested. Finally, the test stub for *Report High Risk Patients* could be replaced with its implementation. Final testing would then verify the operation of the complete program.

7.3 BOTTOM-UP DESIGN

Although top-down design is the most common approach when developing procedural programs, an alternative approach is *bottom-up design*. The philosophy of bottom-up design can be best understood by drawing an analogy with children's building blocks. When building a structure from building blocks we would probably have a general idea of what we wanted to make to begin with (e.g. a house), but it is common to start off by building some small components of the overall structure (e.g. the walls). Once enough of the components are complete, they can be fitted together.

The same principle applies to bottom-up software design: we begin with a general idea of the system to be developed (not specified in detail), and then start off by developing some small self-contained modules. When enough of the modules are complete, they can be combined to form larger subsystems. After each combination step, the developing program can be tested. Like the top-down approach discussed in Section 7.2.1, this can be thought of as a form of incremental development (see Section 4.2): the program is developed module-by-module, with some increments consisting of combining existing modules to form larger modules. This process is repeated until the complete system has been produced.

A key feature of development using bottom-up design is early coding and testing. Whereas, with top-down design, no coding is performed until a certain level of detail is reached in the design, with bottom-up design, coding starts quite early. Therefore, there is a risk that modules may be coded without having a clear idea of how they link to other parts of the system, and that such linking may not be as easy as first thought. On the other hand, a bottom-up approach is well-suited to benefiting from *code reuse*. Code reuse refers to taking advantage of previously developed and tested, modular pieces of code. It is generally considered to be a good thing in software development, since it reduces the amount of duplicated effort in writing the same or similar code modules, and it also reduces the chances of faults being introduced into programs (since the reused modules have been extensively tested already).

To further illustrate the concept of bottom-up design, let's return to our GP surgery risk screening program case study. To develop this program using bottom-up design, we would first consider what types of function/module we might need to write in order to develop a program for risk screening. An initial obvious choice might be a module to find examinations for a given patient. So we would start off by specifying the requirements for this module, designing it using structure charts and/or pseudocode, and then writing the code and testing it. Next, we might choose to expand this module to check if any of the ages at the examinations were greater than or equal to 75, and to build up a list of high risk patients. This module would be specified/designed/coded/tested. Next, we might write a module to check if the list is empty and terminate the program if it is. This new module would be combined with the existing module to make a new module for selecting high risk patients. This process of expanding and combining modules would continue until all modules were completed and tested, forming the final program.

7.4 A COMBINED APPROACH

Although top-down and bottom-up design seem to represent opposing philosophies of software design, most modern software design techniques combine both approaches. Although a thorough understanding of the complete system is usually considered necessary to produce a high quality design, leading theoretically to a more top-down approach, most software projects attempt to maximize code reuse. This reuse of existing code modules introduces more of a bottom-up flavor. You may find it useful to adopt such an approach when designing your own software, i.e. take a predominantly top-down approach to the software design process (e.g. using structure charts and pseudocode), but always keep in mind the possibilities for code reuse. Where possible, try to include modules in your design that can be reused from other programs you have written in the past or from freely available code on the internet.

7.5 ALTERNATIVE DESIGN APPROACHES

In this chapter we have introduced the basics of software design using structure charts and pseudocode. However, in reality software design is a large and complex field and many authors have written entire books on the subject. Alternative design notations and approaches do exist and we will not attempt to cover them all here. Our aim in introducing this topic was not to argue for a particular notation or approach but rather to give a flavor for the field and to illustrate the key point that investing time in designing a program before writing code can be a good idea.

■ Activity 7.1

O7.A, O7.B, O7.C, O7.D

Write a MATLAB implementation of the design we produced for Example 7.1 (i.e. the GP surgery risk screening program). You will need to consider whether to implement each module using a sequence of statements in the main script *m*-file or as a separate MATLAB function. Use an incremental development approach and test stubs when developing your solution. ■

■ Activity 7.2

O7.A, O7.B, O7.C, O7.D

The GP now wishes to expand her definition of a 'high risk' patient by also including those patients who have a body mass index (BMI) of less than 18.5 or greater than 25 at their most recent examination. Recall that a patient's BMI can be calculated using the formula

$$\text{BMI} = \text{mass}/\text{height}^2$$

where the height is specified in meters and the weight in kilograms.

Modify your program design and implementation from Activity 7.1 to meet this new requirement. ■

■ Activity 7.3

O7.A, O7.B, O7.C, O7.D

Now modify your program design and implementation from Activity 7.2 so that the program also reports the reason for a patient being high risk, i.e. either because of their age or their BMI or both. ■

■ Activity 7.4

O7.A, O7.B, O7.C, O7.D

Adjust your program design and implementation from Activity 7.3 so that the program also reports a patient as high risk if their total cholesterol level is greater than 5 mmol/L. Again, the reason for being reported as high risk should be displayed. ■

■ Activity 7.5

O7.A, O7.B, O7.C, O7.D

Finally, modify your program design and implementation from Activity 7.4 so that the program will also consider the patients' systolic and diastolic blood pressures. The rules should be:

- If the systolic blood pressure is $\geq$ 120 mmHg but $<$ 140 mmHg, or the diastolic blood pressure is $\geq$ 80 mmHg but $<$ 90 mmHg, the patient is classified as having *pre-high blood pressure*. These patients should only be high risk if another one of the risk criteria is met. The pre-high blood pressure should also be reported as a reason for being at risk.
- If the systolic blood pressure is $\geq$ 140 mmHg, or the diastolic blood pressure is $\geq$ 90 mmHg, the patient is classified as having *high blood pressure*. These patients should be reported as high risk even if no other criteria are met. ■

7.6 SUMMARY

There are two opposing philosophies to approaching the software design process: *top-down* design and *bottom-up* design.

Top-down design is also known as *stepwise refinement* and involves successively breaking down problems into simpler sub-problems, stopping when the sub-problems are simple enough to tackle on their own. The process of identifying sub-problems and the interactions between them is referred to as *factoring*. Top-down designs can be documented using a graphical technique such as *structure charts*, and/or text-based techniques such as *pseudocode*. Top-down designs can be incrementally developed by making use of *test stubs*, which are simple functions or lines of code that act as a temporary replacement for the full, yet to be implemented code.

Bottom-up design involves starting off with a general idea of the system to be developed, but not documenting it in detail. Rather, small, self-contained modules that are likely to be useful are identified, their requirements specified, a design produced and code written and tested. These low-level modules are successively combined to form higher-level modules. This process continues until the final program has been formed and tested. With bottom-up designs there is a risk that modules may be coded without having a clear idea of how they might link to other parts of the system. On the other hand, bottom-up approaches are well-suited to the concept of *code reuse*, (making use of previously written and tested code modules).

Modern software design approaches generally feature a combination of top-down and bottom-up design: a predominantly top-down approach is taken but care is taken to identify modules that can be reused from previous implementations. As a general rule, for larger programs, it's a good idea not to begin coding straight away. It's almost always a good idea to plan our design on paper first using tools such as structure charts and pseudocode.

7.7 FURTHER RESOURCES

- Top-down vs. bottom-up design: http://en.wikipedia.org/wiki/Top-down_and_bottom-up_design.
- Test stubs: http://en.wikipedia.org/wiki/Test_stub.
- Structure charts: http://en.wikipedia.org/wiki/Structure_chart.

EXERCISES

■ Exercise 7.1

O7.A, O7.B, O7.C, O7.D

In Activity 5.3 we introduced the concept of an *ejection fraction* (EF) in cardiology. Recall that the *end diastolic volume* (EDV) and *end systolic volume* (ESV) of the heart's left ventricle refer to the volume (in mL) of blood just prior to the heart beat and at the end of the heart beat respectively. The EF is the fraction of blood pumped out from the heart in each beat, and can be computed from the EDV and ESV as follows:

$$EF = \frac{EDV - ESV}{EDV}.$$

The normal range of EF values in healthy adults is 55–70%.

The text file *heart_data.txt* contains measured EDV and ESV values for a number of patients. The first few lines of the file are shown below:

```
1111 120 61
2222 136 49
3333 89 64
4444 111 42
5555 160 119
```

The first column represents a patient ID, the second column the EDV and the third column the ESV.

A program should read in these data from the file, compute the EF for each patient, and select only those patients who have an EF outside the normal range. The patient ID, EDV, ESV and EF for only these abnormal patients should then be written to a new text file called *at_risk.txt*. The format of this new file should be as follows:

```
1111 120 61 0.49167
3333 89 64 0.2809
5555 160 119 0.25625
```

Design and implement a program to meet these requirements. Use structure charts and pseudocode in your design, and incremental development/test stubs when writing your implementation. ■

■ Exercise 7.2

O7.A, O7.B, O7.C, O7.D

The *arterial resistivity index* (ARI) is a measure of pulsatile blood flow that indicates the resistance to blood flow through an artery. The ARI is a common way of characterizing the arterial waveform and is used in applications such as detection and monitoring of kidney disease, detection of pre-eclampsia during pregnancy and evaluation of the success of liver transplants. The equation for computing ARI is:

$$ARI = \frac{(v_{systole} - v_{diastole})}{v_{systole}}$$

where $v_{systole}$ and $v_{diastole}$ are the velocities of the blood flow during cardiac *systole* (i.e. contraction) and *diastole* (i.e. relaxation) respectively.

Arterial velocity is commonly measured using *Doppler ultrasound*, which works by measuring the change in frequency of reflected sound waves due to the Doppler effect.

The file *doppler.mat* contains a series of velocity measurements made from the carotid artery of a human volunteer. The file contains two variables:

`doppler` contains the Doppler velocity measurements in cm/s, and `t` contains the times in milliseconds of these measurements. The peaks in the velocity measurements represent the blood flow at systole whilst the troughs represent the flow at diastole.

A program is required to read in these data and display them as a plot similar to that shown in the figure below.

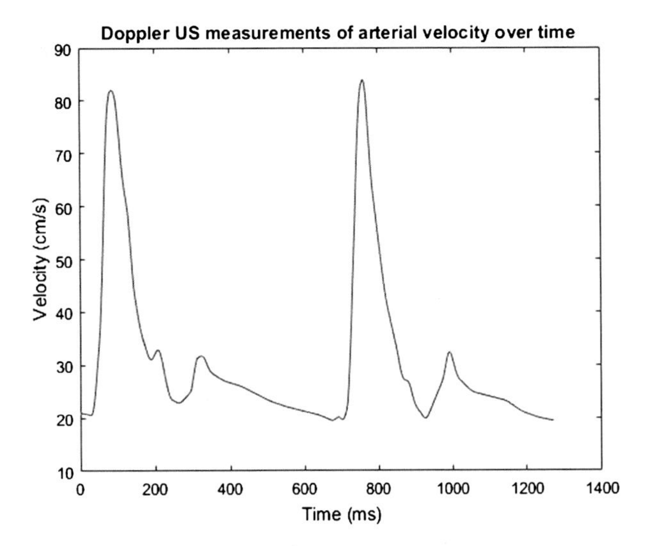

The program should then use the data to compute the ARI. It can be assumed that the systolic velocity is the maximum velocity in the sequence and the diastolic velocity is the minimum.

Design and implement a program to meet these requirements. Use structure charts and pseudocode in your design, and incremental development/test stubs when writing your implementation. ■

FAMOUS COMPUTER PROGRAMMER: DENNIS RITCHIE

Dennis Ritchie was an American computer scientist who was born in 1941 in Bronxville, New York. Whilst a student at Harvard University, he attended a talk on Harvard's computer system and became fascinated with computers and in particular how they could be programmed. As a result he worked in his spare time to learn more and even got a part-time job working at Massachusetts Institute of Technology in their computer labs. In 1968 he received a PhD in Computer Science from Harvard.

Probably Dennis Ritchie's main contribution to the field of computing and computer programming was his early work on operating systems. Operating systems are the software that enable a computer's resources to be accessed by users. In the early days most computers took up entire rooms and their operating systems were very complex and difficult to use. Smaller, portable computers were starting to be developed but Ritchie saw that there was a need for a simpler operating system to bring their power to a wider audience. Therefore, in 1969, he worked with Ken Thompson to develop UNIX. UNIX was much cheaper, simpler and easier to use than any previous operating system. It was hugely influential and is still widely used today.

But this wasn't Ritchie's only major contribution. UNIX was written in machine code, which is specific to the machine it is written for. Ritchie realized that for UNIX to be 'ported' to other machines it must be written in a higher-level programming language, which could be easily compiled on different computer platforms. This led to him making his second major contribution to computing: the invention of the C programming language in 1973. C has proved to be one of the most widely used languages in the history of computer programming, and was also the precursor to C++, another of the most popular programming languages.

Dennis Ritchie died in 2011 at the age of 70 in New Jersey. He died less than a week after the Apple co-founder, Steve Jobs. Whereas Jobs' death was (rightly) met with widespread media coverage and tributes, Ritchie's passing went almost unnoticed in the mainstream media. However, to those in the know, Ritchie's contribution to computing was arguably the greater.

> *"Ritchie was under the radar. His name was not a household name at all, but ... if you had a microscope and could look in a computer, you'd see his work everywhere inside."*
>
> **Paul Ceruzzi**

CHAPTER 8

Visualization

LEARNING OBJECTIVES

At the end of this chapter you should be able to:

O8.A Use MATLAB to visualize multiple plots
O8.B Use MATLAB to produce 3-D plots
O8.C Use MATLAB to read/write/display/manipulate imaging data

8.1 INTRODUCTION

MATLAB is a powerful application for numerical processing and visualization of complex datasets. So far in this book we have seen how to create data using variables (Chapter 1) or to read data from external files (Chapter 6), as well as how to process data using programming constructs (Chapters 2 and 3). In Chapter 1 we also introduced the basics of data visualization in MATLAB. In this chapter we will build on this knowledge and introduce some built-in MATLAB functions for performing more sophisticated visualizations of complex datasets, including images.

8.2 VISUALIZATION

In Chapter 1 we saw how to use the MATLAB `plot` function to produce line plots. We also introduced a number of MATLAB commands to annotate plots produced in this way (e.g. `title`, `xlabel`, `ylabel`, `legend`, etc.). Most of these annotation commands are applicable to other types of plot, as we will see later in this chapter. However, `plot` is only suitable for relatively simple data visualizations, such as visualizing the relationship between two variables or plotting a 1-D signal. In this chapter we will introduce further built-in MATLAB functions that provide more flexibility when producing data visualizations.

MATLAB Programming for Biomedical Engineers and Scientists. DOI: 10.1016/B978-0-12-812203-7.00008-2

8.2.1 Visualizing Multiple Datasets

It is often useful to be able to view multiple visualizations at once, for example to enable us to compare the variations of different variables over time. There are a number of ways in which we can do this using MATLAB.

■ Example 8.1

O8.A

The easiest way to visualize different variables is to carry out multiple plots, either within the same figure or in separate figures. For example, in Section 1.7 we saw how to display multiple plots on the same figure using the `plot` function:

```
x = 0:0.1:2*pi;
y = sin(x);
y2 = cos(x);
plot(x,y,'-b', x,y2,'--r');
title('Sine and cosine curves');
legend('Sine','Cosine');
```

The `plot` function can take arguments in sets of three (i.e. x data, y data, line/marker style). Each set will represent a different line plot in the figure. ■

■ Example 8.2

O8.A

Alternatively, if we have a large number of plots to display, perhaps within a loop in our program, it may be more convenient to make separate calls to the `plot` command to produce our figure. We can do this using the MATLAB `hold` command, as the following example illustrates.

```
x = 0:0.1:2*pi;
y = sin(x);
y2 = cos(x);
plot(x,y,'-b');
hold on;
plot(x,y2,'--r');
title('Sine and cosine curves');
legend('Sine', 'Cosine');
```

The command `hold on` tells MATLAB to display all subsequent plots on the current figure without overwriting its existing contents (the default behavior is to clear the current figure before displaying the new plot). This behavior can be turned off using the command `hold off`. ■

■ Example 8.3

O8.A

Sometimes, displaying too many plots on the same figure can make visualization more difficult. Alternatively, the plots may represent completely different variables on the x and/or y axes which cannot easily be plotted together in a single figure. In such cases we can use multiple figures. We can achieve this in MATLAB through the use of the `figure` command, e.g.

```
x = 0:0.1:2*pi;
y = sin(x);
y2 = cos(x);
plot(x,y);
title('Sine curve');
figure;
plot(x,y2);
title('Cosine curve');
```

Here, the `figure` command tells MATLAB to create a new figure window (keeping any existing figure windows intact) and to make it the current figure. Any subsequent plotting commands will be displayed in the new figure. ■

■ Example 8.4

Creating multiple figures as in the previous examples can be useful, but we might prefer not to generate too many figure windows. An alternative way of generating multiple separate plots is to use subplots within a single figure. The MATLAB `subplot` command allows us to do this, as this example illustrates. *O8.A*

```
x = 0:0.1:2*pi;
y = sin(x);
y2 = cos(x);
subplot(2,1,1);
plot(x,y);
title('Sine curve');
subplot(2,1,2);
plot(x,y2);
title('Cosine curve');
```

This code produces the output shown in Fig. 8.1. From the code shown above, we can see that `subplot` takes three arguments. The first two are the numbers of rows and columns in the rectangular grid of subplots that will be created (in this case, 2 rows and 1 column). The third argument is the number of the current subplot within this grid, i.e. where subsequent plots will be displayed. MATLAB numbers its subplots using the 'row major' convention: the first subplot is at row 1, column 1, the second subplot is at row 1, column 2, and so on. ■

■ Example 8.5

Finally, there is occasionally a requirement to display multiple plots on the same figure with different scales and/or units on their y-axes. The following example illustrates this. This is a program that visualizes two exponentially decaying oscillating functions with very different scales. The output of the program is shown in Fig. 8.2. *O8.A*

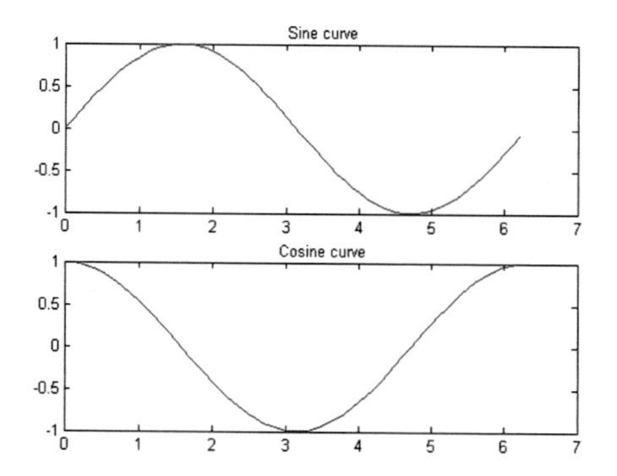

FIGURE 8.1 Example of the use of the `subplot` command to generate multiple plots in a single figure.

```
% generate data
x = 0:0.01:20;
y1 = 200*exp(-0.05*x).*sin(x);
y2 = 0.8*exp(-0.5*x).*sin(10*x);

% produce plots
figure;
yyaxis left;
plot(x,y1);
yyaxis right;
plot(x,y2);

% annotate plots
title('Exponentially decaying oscillations');
xlabel('Time');
yyaxis left;
ylabel('Low frequency oscillation');
yyaxis right;
ylabel('High frequency oscillation');
```

The built-in MATLAB function `yyaxis` is used to display a single plot with two different y-axes: the *left* and *right* axes. The command `yyaxis left` means that all subsequent plots and annotations will relate to the left-hand y-axis, whereas `yyaxis right` activates the right-hand y-axis. Note that the scales of the y-axes are determined automatically by the ranges of the two datasets (`y1` and `y2`). The two datasets share a common x-axis. ■

■ Activity 8.1

O8.A

The file *patient_data.txt* contains four pieces of data for multiple patients: whether they are a smoker or not ('Y'/'N'), their age, their resting heart rate and their systolic blood pressure. Write a program to read in these data and produce separate arrays of age and heart rate data for smokers and non-smokers.

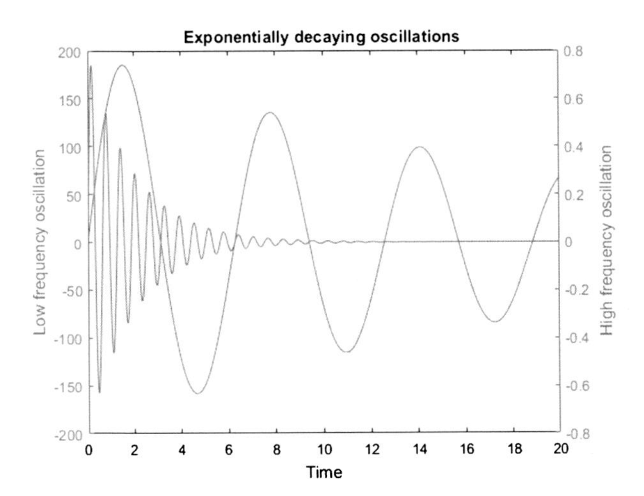

FIGURE 8.2 Example of the use of the `yyaxis` command to generate multiple plots in a single figure with different *y*-axes.

Produce plots of age against heart rate for smokers and for non-smokers,

1. On the same figure with a single use of the `plot` command.
2. On multiple figures using the `figure` command.
3. On the same figure using the `hold` command.
4. On subplots using the `subplot` command. ■

■ Activity 8.2

Electromyography (EMG) involves measurement of muscle activity using electrodes, which are normally attached to the skin surface. One application of EMG is in *gait analysis* (see Activity 1.8). *O8.A*

The file *emg.mat* contains EMG data of the left tibialis anterior muscle acquired from a patient with a neurological disorder that affects movement. The file contains the following variables:

- `times`: The times of the EMG measurements (in seconds).
- `LTA`: The EMG measurements from the left tibialis anterior muscle.

In addition, the *emg.mat* file contains two additional variables representing information about specific gait events that have been identified:

- `foot_off`: An array containing the timings (in seconds) of 'foot off' events (i.e. when the toe leaves the ground).
- `foot_strike`: An array containing the timings (in seconds) of 'foot strike' events (i.e. when the heel hits the ground).

Produce a plot of time (on the *x*-axis) against EMG measurement (on the *y*-axis) and annotate the plot appropriately. In addition, add lines indicating

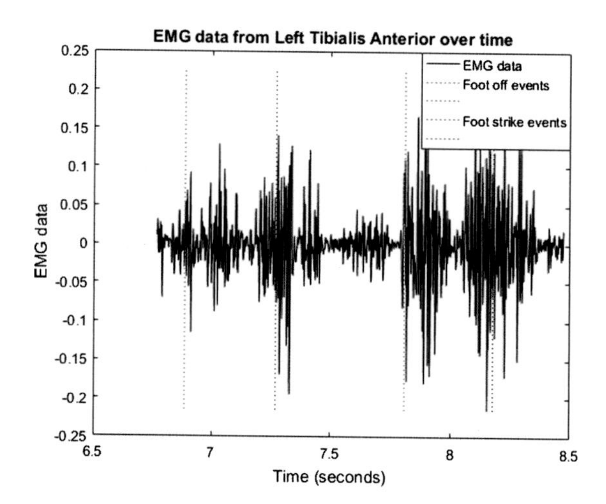

FIGURE 8.3 A plot of EMG measurements showing specific gait events.

the timings of the gait events provided. Add a legend to indicate the meanings of the different lines in your graph. The final figure should look similar to that shown in Fig. 8.3. ■

8.2.2 3-D Plotting

Often when working in biomedical applications it is useful to be able to visualize and analyze data which have several different values per 'individual' datum. For example, we might want to visualize points in 3-D space: in this case the individual data are the points, and each point has three associated values (its x, y and z coordinates). Alternatively, we might have a range of measurements related to hospital patients' health (e.g. blood pressure, cholesterol level, BMI, etc.): in this case the individuals are the patients and there will be multiple measurement values for each patient. This type of data is called *multivariate* data.

■ Example 8.6

O8.B

The first command we will discuss for visualizing multivariate data is the `plot3` function. The following example (adapted from the MATLAB documentation) illustrates the use of `plot3`. This program will display a helix, and its output is shown in Fig. 8.4.

```
% the z-coordinate of the helix
t = 0:0.05:10*pi; % 5 times around a circle

% the x-coordinate of the helix
st = sin(t);
```

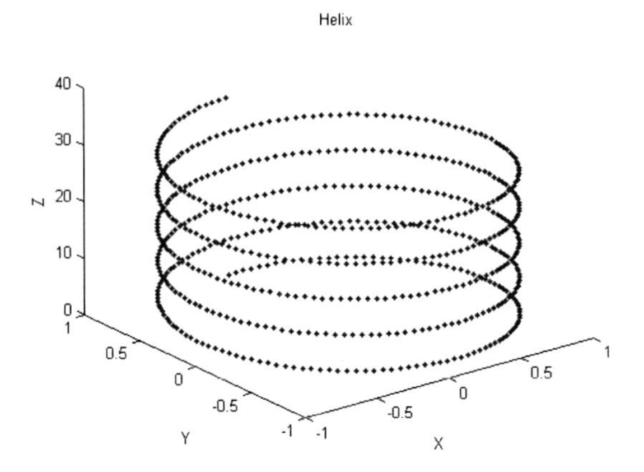

FIGURE 8.4 Output of the helix program shown in Example 8.6.

```
% the y-coordinate of the helix
ct = cos(t);

% 3-D plot
figure;
plot3(st,ct,t, '.b')
title('Helix');
xlabel('X');
ylabel('Y');
zlabel('Z');
```

The first three (non-comment) program statements set up the array variables for storing the x, y and z coordinates of the helix. These three arrays will all have the same number of elements, which corresponds to the number of 3-D points. Each 3-D point will take its coordinates from the corresponding elements in the arrays. The `plot3` function takes four arguments: the first three are the arrays of coordinates, and the fourth (optional) argument specifies the line/marker appearance, i.e. the same as for the `plot` function (see Section 1.7). Note that the same annotation commands (`title`, `xlabel`, etc.) can be used with figures produced by `plot3`, but that now we have an extra axis to annotate using `zlabel`. ■

■ Activity 8.3

O8.B

Using the same *patient_data.txt* file you used in Activity 8.1, write a MATLAB program to produce a 3-D plot of the patient data. The three coordinates of the plot should be the age, heart rate and blood pressure of the patients, and the smokers and non-smokers should be displayed using different symbols on the same plot. Annotate your figure appropriately by, for example, inserting suitable labels on the axes. ■

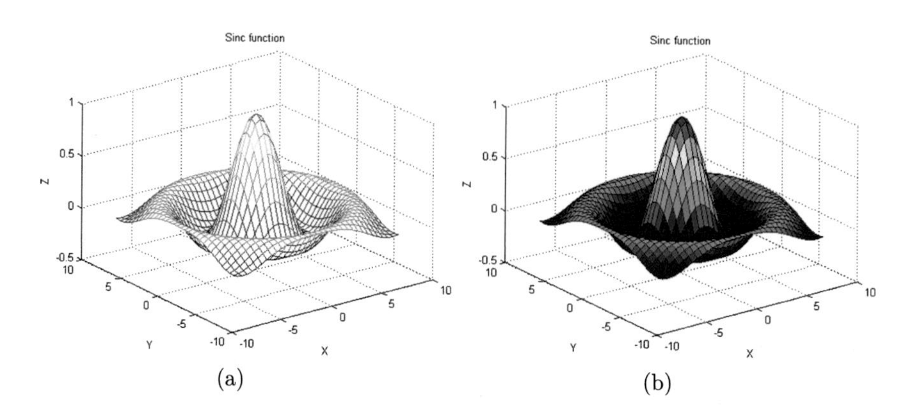

FIGURE 8.5 **Output of the sinc function program shown in Example 8.7: (a) using the `mesh` command; (b) using the `surf` command.**

MATLAB also provides two built-in functions for visualizing multivariate data as *surfaces*: `mesh` and `surf`. An example of such a visualization is shown in Fig. 8.5a, which is the output of the program shown in Example 8.7 below. This type of plot is commonly used for visualizing data in which there is a single z value for each point on an x–y plane. (Note that this isn't the case for the helix data in Example 8.6 as a given pair of x, y values can have multiple z values.)

■ Example 8.7

O8.B

In this example (adapted from the MATLAB documentation) a mesh plot is produced to illustrate the *sinc* function (i.e. $\sin(x)/x$).

```
% create arrays representing a grid of x/y values
[X,Y] = meshgrid(−8:.5:8, −8:.5:8);

% define sinc function
R = sqrt(X.^2 + Y.^2) + eps;
Z = sin(R)./R;

% mesh plot
mesh(X,Y,Z)
%surf(X,Y,Z)
title('Sinc function');
xlabel('X');
ylabel('Y');
zlabel('Z');
```

The built-in MATLAB `meshgrid` command creates two rectangular arrays of values to define the x–y grid over which the mesh is plotted. One array indicates the x-coordinate at each grid point and the corresponding element in the other array indicates the grid point's y-coordinate. See Section 8.2.3

for further details of the `meshgrid` command. Arrays of this form, together with a corresponding array of z values, are the arguments required by the `mesh` command.

The values calculated for the z-coordinate array are the distances from the origin of each grid point, after applying the *sinc* function. Note the use of the `.^` operator to square the x and y coordinates: this is necessary to ensure that the `^` operator is applied element-wise to the array (rather than performing a *matrix power* operation, see Section 1.4). The `eps` constant in MATLAB returns a very small floating point number and is necessary to avoid a division-by-zero error at the origin of the x–y plane.

The `surf` command has the same form as `mesh`, as we can see from the commented-out command in the code above. The result of using `surf` is to produce a filled surface rather than a wireframe mesh (see Fig. 8.5b). ■

8.2.3 The `meshgrid` Command

The `meshgrid` command introduced in the previous example is useful for 3-D plotting of gridded data. This section describes in more detail how it works. As we saw, `meshgrid` generates arrays containing the coordinate values of all points on a regular grid. The syntax for making a 2-D grid is

```
[X,Y] = meshgrid(x, y)
```

The input arguments are two 1-D vectors containing the coordinate values required along each dimension: one for x and one for y. In this case, the output arguments, `X` and `Y`, are 2-D arrays with the same shape as the grid, containing *all* the grid points' x and y coordinates. Each one is generated by replicating the input 1-D vectors, horizontally or vertically. An example call to make a small grid is as follows:

```
>> [X, Y] = meshgrid(1:4, 2:3)

X =
     1     2     3     4
     1     2     3     4


Y =
     2     2     2     2
     3     3     3     3
```

Note that the y coordinates increase as we read *down* the array because MATLAB arrays are indexed starting from the top left. This is equivalent to viewing the x–y axes as shown in Fig. 8.6, which illustrates how the matrices `X` and `Y` in the above call are generated.

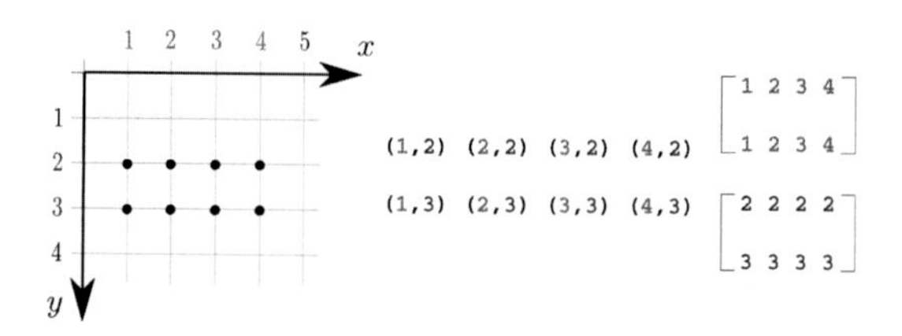

FIGURE 8.6 An illustration of the use of the `meshgrid` command. The grid required is shown by the black dots on the left-hand axes (with *y* pointing down). The grid ranges from $x = 1$ to $x = 4$ and $y = 2$ to $y = 3$. The coordinates for all required grid points are shown in the middle of the figure, with *x*-coordinates shown in red and *y*-coordinates shown in blue. The outputs from `[X,Y] = meshgrid(1:4, 2:3)` are two separate arrays shown on the right of the figure, one containing the grid *x* coordinates and one containing the grid *y* coordinates.

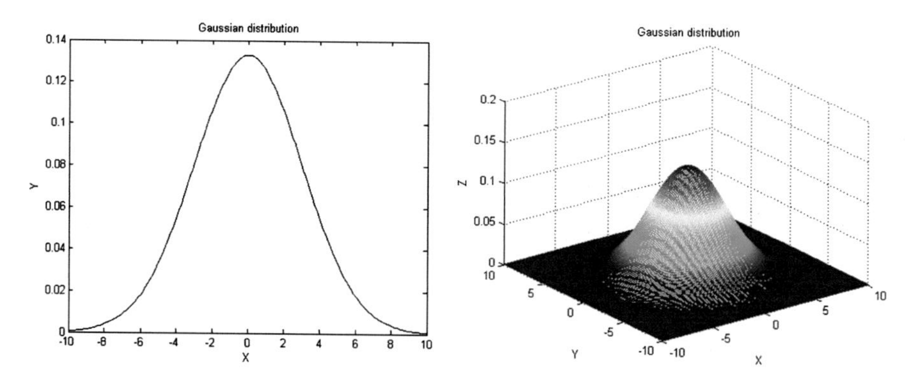

FIGURE 8.7 1-D and 2-D Gaussian distributions.

■ Activity 8.4

O8.A, O8.B

A (1-D) Gaussian (or *normal*) distribution is defined by the following equation:

$$\frac{1}{\sigma\sqrt{2\pi}}e^{\frac{-(x-\mu)^2}{2\sigma^2}} \tag{8.1}$$

where x is the distance from the mean, μ, of the distribution. The parameter σ is the standard deviation of the distribution, which affects how 'spread out' the distribution will be. An example of such a distribution is shown in Fig. 8.7 (left).

1. Write a MATLAB function to display a figure of a 1-D Gaussian distribution. The function should take the following arguments: the standard deviation and the mean, and the minimum/maximum x axis values for the plot. You can experiment with different standard deviations but to begin with try a value of 3 with a mean of 0 and plot the distribution between x coordinates of -10 and $+10$.

2. Write another MATLAB function to display a 2-D Gaussian distribution. To calculate the function for a 2-D point, you should replace the $(x - \mu)$ term in Eq. (8.1) with the distance of the point from the center of the distribution (for example, the origin, $(0, 0)$). See Fig. 8.7 (right) for an example. This function should take the following arguments: the standard deviation of the distribution, the x/y coordinates of the center of the distribution, and the minimum/maximum axis values. You can experiment with different standard deviations but to begin with try a value of 3 and plot the distribution between the x/y coordinates ranging from -10 to $+10$, and use a center of $(0, 0)$. ■

8.2.4 Imaging Data

The use of images in biomedical engineering is increasingly common. Medical images can be acquired using modalities such as magnetic resonance (MR), X-ray computed tomography (CT) and ultrasound (US). In this section we will introduce how to read, write and display imaging data using MATLAB. Chapter 10 will deal with images in MATLAB in more detail.

■ Example 8.8

The following example illustrates the use of several built-in image-related MATLAB commands. *O8.C*

```
% read image
im = imread('fluoro.tif');

% display image
imshow(im);

% modify image
im(100:120,100:120) = 0;

% save image (saves 'im' array)
imwrite(im, 'fluoro_imwrite.tif');

% save image (saves figure as displayed)
saveas(gcf, 'fluoro_saveas.tif');
```

The `imread` command reads imaging data from a file with a name specified by the single argument. The value returned is a 2-D array of pixel intensities. The *fluoro.tif* file is a *fluoroscopy* image (i.e. a low dose real-time X-ray) and is available from the book's web site.

Images are treated as normal 2-D arrays in MATLAB. The intensities can be of different types such as `int8`, `single` or `double`, but within a single image all must have the same type, as with all MATLAB arrays. Imaging data can be manipulated just like any other MATLAB array, as we can see from the third non-comment line in the above example in which a rectangular block of pixels has its intensities set to 0.

The `imshow` command can be used to display a 2-D image in a figure window. The single argument is the 2-D array of pixel intensities.

This example shows two ways in which the image can be written to an external file. The `imwrite` command takes a 2-D array and a file name as its only arguments, so the image data written to the file depends only on the array and not on any figure currently being displayed. The `saveas` command is used in this example to save the data in the current figure (`gcf` means *get current figure*). Therefore, any changes made to the displayed figure will also be saved, but any changes made to the original array since it was displayed (as in the example above) will not be saved. In this example, `saveas` saves the figure as an image, but it can also be used to save in the native MATLAB figure format, which has a ".fig" file extension. ■

Both the `imread` and `imwrite` commands can handle a range of different imaging formats, such as `bmp`, `png` and `tif`. For a full list of supported formats, type `imformats` at the MATLAB command window. Normally, just adding the extension to the file name is enough to tell MATLAB which format we want to use.

In medical imaging, it is common for images to be 3-D datasets rather than 2-D (e.g. many MR and CT images and some US images). Unfortunately, MATLAB doesn't have any built-in functions for visualizing 3-D images. However, several have been written by the MATLAB user community and been made available for free download from the Mathworks File Exchange web site. For example, the `imshow3Dfull` function allows simple interactive slice-by-slice visualization of 3-D image datasets. See the Further Resources section at the end of this chapter for details.

■ Activity 8.5

O8.C

Digital subtraction angiography is an imaging technique in which two fluoroscopy images are acquired: one before injection of a contrast agent and one shortly after. The pre-contrast image (the *mask* image) is subtracted from the contrast enhanced image (the *live* image) to provide enhanced visualization of blood vessels.

Write a MATLAB script to:

- Ask a user to choose mask and live image files using a file selection dialog box.*
- Load in the mask and live image files.
- Subtract the mask from the live image.
- Display the resulting digital subtraction angiography image.

Sample mask and live images (*dsa_mask.tif* and *dsa_live.tif*) are available for download from the book's web site.

*(*Hint: Look at the MATLAB documentation for the* `uigetfile` *function.)* ■

8.3 SUMMARY

MATLAB provides a number of built-in functions for performing more sophisticated data visualizations. The `plot`, `hold`, `figure`, `subplot` and `yyaxis` commands can be used to visualize multiple datasets at the same time. 3-D plots can be produced using the `plot3`, `mesh` and `surf` commands. 2-D grids of coordinate values suitable for use with `mesh` and `surf` can be generated using the `meshgrid` command.

Imaging data can be read and assigned to 2-D array variables using `imread`, and displayed using `imshow`. An array can be saved to an image file using `imwrite`, and a figure that is currently being displayed can be saved using `saveas`. A number of 3-D image visualization tools are available for free download from the Mathworks File Exchange web site.

We will return to the topic of data visualization, in the context of statistical analysis, in Chapter 12.

8.4 FURTHER RESOURCES

- MATLAB documentation on 2-D and 3-D plots: http://www.mathworks.co.uk/help/matlab/2-and-3d-plots.html.
- MATLAB documentation on image file operations: http://www.mathworks.co.uk/help/matlab/image-file-operations.html.
- Mathworks File Exchange: http://www.mathworks.co.uk/matlabcentral/fileexchange.
- `imshow3Dfull` function: http://www.mathworks.co.uk/matlabcentral/fileexchange/47463-imshow3dfull–3d-imshow-in-3-views-.

EXERCISES

■ Exercise 8.1

In Exercise 4.4 we introduced data measured using an oscillatory blood pressure monitoring device. The data (contained in the file *blood_pressure.mat*) comprise an array (`p`) of pressure measurements in mmHg and an array (`peaks`) containing zeros and ones, with a value of one where peaks in the pressure oscillations were identified (zero otherwise). Produce a plot of both datasets in the same figure but with two different y axes for the two arrays. Annotate your figure appropriately. ■

O8.A

■ Exercise 8.2

The process of aligning 3-D medical images is known as *image registration*. When assessing how well two images have been registered (aligned), it is possible to use a variety of measures. A research team evaluates the registration of a number of image pairs using two different measures. The first is

O8.A

the value of the *normalized cross correlation* (a measure of image similarity) between each image pair. The second is the *landmark error*, which is determined by observers manually placing corresponding anatomical landmarks in each image pair and computing the distance between them after registration. A good registration should give a lower landmark error. The values of these measures for 10 separate experiments are provided in the files *ncc.txt* and *points.txt*. Both files contain 10 rows arranged in two columns that represent the experiment number and the respective evaluation measure.

Write a MATLAB program to read in these data and produce a single plot showing the values of both measures (on the *y*-axis) against the experiment number (on the *x*-axis). Annotate the *x* and *y* axes of your plot and add a suitable title. Your final figure should look like the one shown below.

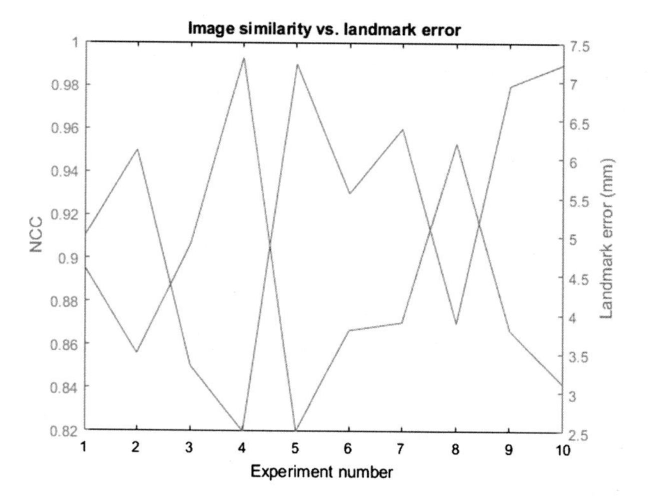

■

■ Exercise 8.3

O8.A

This exercise continues the case study that we have already seen in Example 1.2 and Example 4.1. Recall that we have data representing radial displacements of the left ventricular myocardium of the heart. We have these data separately for the 17 AHA segments. Now we wish to visualize the displacements for all segments in a single plot.

The input to your program will be a file containing radial displacement data broken down by AHA segment. The basic format of the file is:

```
tres
s1t1 s1t2 ... s1tR
s2t1 s2t2 ... s2tR
...
s17t1 s17t2 ... s17tR
```

where `tres` represents the *temporal resolution* of the radial displacement data, i.e. the amount of time (in milliseconds) between subsequent displacement measurements. Each other entry in the file represents a radial displacement measurement for a particular segment (`s`) at a particular time (`t`) in the cardiac cycle. `R` is the number of radial displacements recorded in a particular cardiac cycle. For the purposes of this exercise, you can assume that `R` will always be equal to 20, and there will always be 17 segments.

The following sample input files are provided:

- *radial_motion_aha_pat_01.txt*
- *radial_motion_aha_pat_02.txt*
- *radial_motion_aha_pat_03.txt*

Your task is to write a MATLAB script that reads in the data from a specified file and displays a single figure that plots the radial displacement for each AHA segment against time in milliseconds. The output of your script for the first sample input file is shown in the figure below.

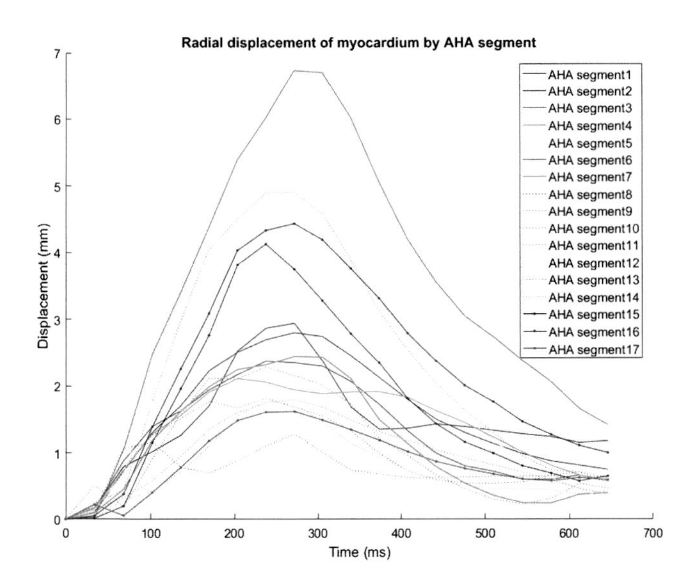

Note the following features of this figure that your code should reproduce:

- The lines for each segment should be displayed using a different style/color.
- A legend should be included indicating which line corresponds to which segment (in the input file, the data for segment 1 are first, segment 2 second, and so on).

Try to use an iteration statement in your solution. ■

■ Exercise 8.4

O8.A

In Activity 1.8 we introduced knee flexion data that were used in gait analysis. The file *joint_flexion.mat* contains the same knee data as well as data from the hip and ankle. Specifically, the file contains the following variables:

- `times`: The timings of the flexion values (in seconds).
- `angles`: A matrix in which the first column represents flexion (i.e. angle) of the left hip, the second column is flexion of the left knee and the third column is flexion of the left ankle.
- `foot_off`: An array containing the timings (in seconds) of 'foot off' events (i.e. when the toe leaves the ground).
- `foot_strike`: An array containing the timings (in seconds) of 'foot strike' events (i.e. when the heel hits the ground).

Produce a plot of joint flexion against time for all three joints on the same figure and annotate the plot appropriately. As in Activity 1.8, add lines indicating the timings of the 'foot off' and 'foot strike' events and a legend indicating the meanings of the different lines. ■

■ Exercise 8.5

O8.B

Optical tracking systems have a range of applications in biomedical engineering, from gait analysis to image guided surgery. This technology typically uses a pair of cameras which localize markers attached to the object being tracked. The 3-D location of a marker can be determined from a pair of 2-D image pixel locations using triangulation.

A company are developing a new optical tracking system and wish to evaluate its tracking accuracy. They have tested the system on 'phantom' data (i.e. data acquired from a specially designed test object), and have collected 3-D localization error data. The data consist of errors along the x, y and z axes for 50 separate tests, where the x/y axes are parallel to the image plane of one of the cameras, and the z axis is the direction of the same camera.

These error data are available in the file *optical_tracking.mat*. Write a MATLAB script to produce a 3-D plot of the x, y and z errors, to enable visualization of the directional dependency of the errors. What do you observe? Can you offer an explanation? ■

■ Exercise 8.6

O8.B

In Exercise 8.2 we introduced the concept of *image registration*. When evaluating the behavior of image registration algorithms it can be useful to visualize the 'similarity landscape' of a pair of images. This means we evaluate a similarity measure at a number of different registration parameters, and then visualize the variation of the similarity measure as a function of the registration parameters.

Suppose that we have done this for a pair of images. x and y translations ranging from -10 mm to $+10$ mm have been applied to one of the images, and the resulting normalized cross correlation (NCC) similarity measure has been computed. These data are available to you in the file *similarity_measure.mat*. The file contains a 21×21 array, representing the NCC value for x and y translations from -10 mm to $+10$ mm in steps of 1 mm.

First, write a MATLAB script to visualize the NCC data as a function of x and y translation. Annotate your plot appropriately. Next, write MATLAB code to find the peak NCC value and the corresponding x and y translations.

(Hint: Look at the MATLAB documentation for the `ind2sub` *command.)* ■

■ Exercise 8.7

O8.C

Write a MATLAB program that will read in an image from a file and display it, and then allow the user to *annotate* the image (i.e. add text strings to it at specified locations). To perform the annotation the user should be prompted to enter a string, then click their mouse in the image to indicate where they want the string to appear. The user should be allowed to add multiple annotations in this way, and the program should terminate and save the annotated image with a new file name when an empty string is entered.

A sample image file, called *brain_mr.tif*, which you can use to test your program, is available through the book's web site.

(Hint: Look at the MATLAB documentation for the `gtext` *command.)* ■

■ Exercise 8.8

O8.C

Try visualizing 3-D imaging data using the `imshow3Dfull` function. The MATLAB code for `imshow3Dfull` can be downloaded from the Mathworks File Exchange (see Further Resources section in this chapter). Sample 3-D imaging data can be downloaded from the book's web site as the *MAT* file *MR_heart.mat*. ■

FAMOUS COMPUTER PROGRAMMER: TIM BERNERS-LEE

Tim Berners-Lee is a British computer scientist who is best known as the inventor of the World-Wide-Web (WWW). He was born in 1955 in London, UK. His parents worked on the first commercially built computer, the Ferranti Mark 1. Tim was a keen trainspotter as a child, and he learned about electronics from playing with his model railway. Later, he went on to study Physics at Queen's College, Oxford, graduating in 1976.

From 1980 he worked as a computer programmer at the European Nuclear Research Lab, CERN, in Geneva, Switzerland. Whilst there he wrote a computer program called ENQUIRE, which was a piece of software for storing information together with associations between them – in writing this program Berners-Lee had actually invented the concept of 'hypertext' that is so common on the WWW today. After working in the UK for a few years, in 1984 he returned to CERN. In 1989 he saw an opportunity to utilize the concept of hypertext to make the internet more accessible. The result was the WWW. In his own words, "creating the web was really an act of desperation, because the situation without it was very difficult". He wrote software for the first WWW server, and developed specifications for HTML and HTTP, which are the basis for almost all web sites in the world today. Throughout, he has made his work available freely without patent, and has earned no royalties from the success of the WWW.

In 1994 he founded the World Wide Web Consortium (W3C), the not-for-profit organization that coordinates the development of the WWW. He received a knighthood from the UK government in 2004, and in 2012 he was a part of the London Olympics opening ceremony, which celebrated his immense and selfless contribution to the development of the internet as we know it today, all of which was made possible by computer programming.

He is currently a Professor at the University of Southampton in the UK and works on a variety of projects related to the WWW and net neutrality.

> *"When somebody has learned how to program a computer ... You're joining a group of people who can do incredible things. They can make the computer do anything they can imagine."*
>
> **Tim Berners-Lee**

CHAPTER 9

Code Efficiency

LEARNING OBJECTIVES

At the end of this chapter you should be able to:

- *O.9A* Explain the importance of time efficiency (speed) for running programs
- *O.9B* Explain the importance of memory efficiency (space) for running programs
- *O.9C* Describe the importance of a program running correctly, as opposed to efficiently
- *O.9D* Describe why a naive implementation of a recursive function can lead to bad performance
- *O.9E* Explain the fundamental ideas of dynamic programming that aim to improve the performance of recursive programs

9.1 INTRODUCTION

When code becomes reasonably long or complex, it may become necessary to consider how efficient it is. There are two senses in which we can talk about efficiency of code: in terms of the time it takes to run and in terms of the amount of computer memory it needs when running. In some cases, a program needs to be very time efficient and should run and produce results quickly, for example in software that is used by cars to manage the running of the engine or software used for instrument control in robotic surgery. In other cases, a program may need to be very economical with the amount of memory it uses, for example in programs running on devices where memory is very limited such as an embedded processor in a domestic appliance or in a wearable ECG monitor. This chapter will consider some aspects of efficiency, focusing on MATLAB specific examples, although some of the issues raised will apply to most languages.

9.2 TIME AND MEMORY EFFICIENCY

As a rule of thumb, in order to assess the memory efficiency of a piece of code, we need to focus on the variables it has, in particular on the largest arrays that it

MATLAB Programming for Biomedical Engineers and Scientists. DOI: 10.1016/B978-0-12-812203-7.00009-4

uses. Recall that MATLAB defaults to double precision when assigning numeric values (see Section 5.2.2) which means we need eight bytes per element in a numeric array (unless we override the default behavior).

To assess how time efficient a piece of code is, we need to break it down into *fundamental instructions*. We make the simplifying assumption that these all roughly take the same amount of time each. Fundamental operations or instructions can include:

- Assigning a value to a variable, e.g. `a = 7;`
- Looking up a value of a particular element in an array, e.g. `myArray(3)`
- Comparing a pair of values, e.g. `if (a > 10) ...`
- Arithmetic operations like adding and multiplying, e.g. `a * 4`

■ Example 9.1

O9.A

We consider a very simple example to illustrate some aspects of code efficiency. The following is a trivial function that takes one argument (`N`) which is treated as an integer and adds up all the integers from 1 up to `N` and reports the result.

```
function sumOfIntegers(N)
% Usage: sumOfIntegers(N)
% Add up all the integers from 1 up to N and report
% the result.

a = 1:N;
total = 0;

for j = 1:N
  total = total + a(j);
end

fprintf('Sum of first %u integers = %u \n', N, total);
```

To keep things simple, we consider the first two assignments (to the variables `a` and `total`) as one assignment each.[1] Then there is a `for` loop which has `N` iterations. Within each iteration there is

- An array look-up: `a(j)`
- A basic arithmetic operation: `total+a(j)`
- An assignment: `total=total + a(j);`

Therefore, we have three operations per iteration which gives a total of $3N$ operations for the entire loop. Counting in this way, the estimate is that the function will need $3N + 2$ fundamental operations. As we can see, the number of operations that the function will carry out depends upon the

---

[1] Really, the assignment to variable `a` is an array assignment so requires allocating more memory than for a single numeric scalar value. This can add a time overhead which we ignore here.

number passed in as the argument N. For large values of N, more operations are carried out. ■

9.2.1 Timing Commands in MATLAB

Recall that the built-in commands `tic` and `toc` can be used to measure execution time (see Section 2.7). They can be used as follows:

```
tic
% some
% commands
% here
toc
```

If we just have a single command, we can use commas or semi-colons to separate them, for example:

```
>> tic; a = sin( exp(3) ); toc
```

or, using commas

```
>> tic, a = sin( exp(3) ), toc
```

The output for the second command could be as follows:

```
a =

    0.9445

Elapsed time is 0.000355 seconds.
```

MATLAB reports that the above example took 355 microseconds. The same example, repeated on a different machine, is likely to run in a different amount of time, due to differences in the hardware and the operating system. Even repeating the same operation on the same machine can lead to different times being reported because the load on the operating system can change between runs.

■ Example 9.2

Returning to the `sumOfIntegers` function from Example 9.1, we can measure how long the function takes to execute for various values of N. Let us try values of one thousand and one million for N. We'll focus on the main loop in the function. We can add a `tic`/`toc` pair around the loop so it becomes

O9.A

```
% ...
tic
for j = 1:N
  total = total + a(j);
```

```
end
toc
% ...
```

■

■ Activity 9.1

O9.A

Download the code for the modified function described in Example 9.2. It is called *sumOfIntegers_2.m*. Run it in the command window for different values of N. In particular, run it for $N = 1000$ and for $N = 1{,}000{,}000$ and make a note of the time taken – you should confirm that the second run takes approximately 1000 times as long as the first. ■

The ratio in the times given will not be exactly 1 to 1000 for a number of reasons, perhaps due to variations in the demands on the operating system between runs or because MATLAB might optimize the code in some way.

9.2.2 Assessing Memory Efficiency

Recall that the `whos` function identifies which variables are available in the workspace and how much memory they are using (see Section 1.5). This can be particularly useful when entering commands that define variables in the command window.

When writing a function of our own, however, we need to adopt a different strategy. We can inspect the code directly to find which variables will require the most memory. In the case of the function `sumOfIntegers` in Example 9.1, the variable that requires the most memory is the variable `a` declared in the command `a = 1:N;` . This statement declares and assigns an array of length `N` when the function is run. Therefore, the amount of memory used by the function depends upon the argument given to it. Larger values of `N` will mean that more memory is needed.

Another way to inspect the amount of memory used by a function is more direct. We can run the function and use the debugger to stop the execution and look directly at how much memory is being used. For example, we can place a break point on the last line in `sumOfIntegers` and call it with a value of one thousand:

```
>> sumOfIntegers(1000)
```

When the code stops on the last line, a call to `whos` gives the following output:

```
K>> whos
  Name        Size            Bytes  Class     Attributes

  N           1x1                 8  double
  a           1x1000           8000  double
```

```
j           1x1                      8  double
total       1x1                      8  double
```

This lists the four variables (`N`, `a`, `j`, and `total`) and how many bytes each one uses. Note that the variables listed are only those within the scope of the function as it is currently being run, and any workspace variables, for example, will not be listed (see Section 3.9).

■ Example 9.3

Returning once again to the trivial function `sumOfIntegers`, an obvious way to avoid a memory overhead is to simply avoid the allocation of the array. For the example, this is quite simple: the code can be re-written as:

O9.B

```
function sumOfIntegers(N)
% ... usage
% ..
total = 0;

for j = 1:N
  total = total + j;
end
% ...
```

■

■ Activity 9.2

Download the original code from Example 9.1 and the code for the modified function described in Example 9.3. They are called *sumOfIntegers.m* and *sumOfIntegers_3.m* respectively. Call each one from the command window using the same value of N. For each one, place a breakpoint on the last line so that the running halts there in the debugger. When the code stops at this breakpoint, for each version, type `whos` in the command window. Use this to confirm that the new version is more memory efficient than the first. ■

O9.B

The above example illustrates an easy win – the code did not need an array at all. But not all examples are as straightforward as this.

It is possible to make excessive memory demands unintentionally. The built-in `ones` or `zeros` functions are commonly used to quickly initialize an array containing only 1's or 0's. By default, when only one argument is given to these functions, it is assumed that the user requires a square matrix (i.e. a 2-D array).

■ Example 9.4 (A *very* memory inefficient piece of code)

If we want a long *vector* of zeros, i.e. a 1-D array, say a vector that has a length of 100,000, it is very easy to make a mistake by calling the `zeros` function as follows:

O9.B

```
myVector =  zeros(100000)  % Bad! do not do this...
```

MATLAB interprets this as a request for a *square matrix* of size $100{,}000 \times 100{,}000$. Therefore, it tries to allocate enough memory for $100{,}000 \times 100{,}000 = 10^{10}$ `double` values. Each `double` value requires eight bytes because that is the default numeric precision in MATLAB. This means that the user is asking for 80,000,000,000 bytes which is a little over 74 GB because

$$74 \times \underset{\text{bytes/KB}}{1024} \times \underset{\text{KB/MB}}{1024} \times \underset{\text{MB/GB}}{1024} = 79456894976 \text{ bytes}$$

Most ordinary computers do not have 74 GB of RAM available, so typing such a command into MATLAB is very inadvisable and likely to cause a crash. ■

■ Activity 9.3

O9.B

Assign to an array using the functions `ones` or `zeros`, using reasonable values for the size of the array, e.g.

```
>> a = zeros(100);
```

and confirm that the variables in the workspace have the size you expect. Carry out the following commands

```
>> b = ones(80);

>> c = ones(80, 80);

>> d = ones(80, 1);
```

and decide which variables take up the same amount of space in memory. ■

9.3 TIPS FOR IMPROVING TIME EFFICIENCY

A couple of tips for improving time efficiency are given below. These are to pre-allocate arrays before using them, and to avoid loops by using vector style operations.

9.3.1 Pre-Allocating Arrays

MATLAB is fairly relaxed about the length of an array and will adjust it automatically if the user decides to assign a value to an element that is beyond its current limits. For example, the command:

```
x = randi(3, 1, 6)
```

will generate a length 1×6 row vector of random integers drawn from the set $\{1, 2, 3\}$, for example $[3, 1, 3, 1, 2, 2]$. If we make an assignment beyond the end of the array, e.g. `x(10)= 5` then it is automatically resized to a size of 1×10. The 'gap' between the end of the original array and the new array will be filled with zeros. This means the resulting vector will become $[3, 1, 3, 1, 2, 2, 0, 0, 0, 5]$.

As discussed in Section 2.7, this can be convenient and enables the user to avoid worrying about whether the element being written to is within the array (yet) or not. This is, however, a costly operation in terms of time. There is an overhead involved in resizing an array and this is made worse if it is done within a loop.

The following code snippet illustrates this. It calculates the first 10 cubic numbers:

```
% Start off with an empty array
cubes = [];
for j = 1:10
  cubes(j) = j^3;
end
```

The code starts with an empty array. Then, in the first iteration, it assigns to the element at index 1 (which doesn't exist yet). So the array is resized to accommodate this element. The second iteration assigns to the element at index 2, and again the array needs to be resized, etc. In each iteration, we are assigning a value to an element in the array `cubes` whose index is one more than its current size. Every iteration leads to a resizing of the array and each resize has an associated cost.

As we saw in Section 2.7, we can avoid this cost by *pre-allocating* the array to achieve much better time efficiency:

```
cubes = zeros(1, 10);
for j = 1:10
  cubes(j) = j^3;
end
```

9.3.2 Avoiding Loops

Among the most time-consuming parts of a program are loops such as `for` or `while` loops. This is especially true if the loops are *nested*, for example:

```
for j = 1:N
  for k = 1:M
    % Some code carried out here
  end
end
```

In this case, the code within the central loop is executed $M \times N$ times (unless a `break` command is contained in the central section of code and can be reached). In other words, nesting loops has a multiplicative effect on the number of operations needed. This is why nesting loops should only be done when it is unavoidable, especially if three or more loops are nested.

One way to avoid loops is by using built-in functions. In the previous version of the `sumOfIntegers` function (Example 9.3), we can replace the entire loop

```
for j = 1:N
...
end
```

with a simple call to the built-in function `sum`:

```
total = sum(1:N);
```

Here, the array initialization code `1:N` is used directly as an input argument to the `sum` function (without actually assigning it to an array). This single line will run faster than the original loop. In fact, many built-in functions can accept array arguments, e.g. `sin`, `exp`, etc., and we can use this fact to avoid unnecessary loops.

■ Example 9.5

O9.A, O9.B

The code below constructs fifty points on the graph $y = \sin(x)$ between $x = 0$ and $x = 2\pi$.

```
x = linspace(0, 2*pi, 50);
y = zeros(size(x(j)));
for j = 1:50
  y(j) = sin(x);
end
```

It tries to improve efficiency by pre-allocating the array `y` to be the same size as `x`. However, since the built-in `sin` function in MATLAB accepts an array argument, we can simply replace the loop with a one-line command to generate the array variable `y` directly. This is done by applying the `sin` function element-wise to array `x`:

```
x = linspace(0, 2*pi, 50);
y = sin(x);
```

■

Operating directly on arrays in this way, in order to avoid loops, is often called *vectorization* or *vectorized calculation*. It can also be applied when we have multiple arrays to process rather than just one as in the previous example.

■ Example 9.6

Suppose we have collected the widths, lengths and heights of four cuboids and we want to calculate their volumes. Say the dimensions are collected into three arrays: O9.A

```
w = [1, 2, 2, 8];
l = [2, 3, 5, 3];
h = [1, 3, 2, 4];
```

We could use a loop to find the volumes:

```
v = zeros(1,4);
for j = 1:4
  v(j) = w(j) * l(j) * h(j);
end
disp(v)
```

which gives the output:

```
     2    18    20    96
```

However, this code can be speeded up by *vectorization*. We use the dot operator to perform element-wise multiplication of our dimension data:

```
v = w .* l .* h;
```

achieving the same result with improved time efficiency. ■

Remember, the above is a 'toy' example (there are only 4 data points in it for a start) so the speed-up will be slight. But if the data arrays were much larger then the time saved would be more significant.

■ Activity 9.4

Re-write the following code so that it avoids using a loop. O9.A, O9.B

```
N = 10;

a = zeros(1, N);
b = zeros(1, N);

for i = 1:N
  a(i) = i;
  b(i) = a(i) * a(i);
end
```

■

■ Example 9.7

Recall the that surface plot of the sinc function that we saw in Example 8.7 O9.A relied on the generation of two arrays `X` and `Y` using the `meshgrid` function and then calculating the height of the function in an array `Z`. The code used vectorization and is repeated here:

```
% create arrays representing a grid of x/y values
[X,Y] = meshgrid(-8:.5:8, -8:.5:8);

% define sinc function in a good way
R = sqrt(X.^2 + Y.^2) + eps;
Z = sin(R)./R;
```

The input arrays are both 33 × 33 arrays and the resulting `Z` array has the same size because the element-wise operations ensure this. ■

■ Activity 9.5

O9.A

Re-write the code from Example 9.7 so that it uses nested loops instead. Which of the versions, with and without loops, is more readable? Which version is most efficient? ■

9.3.3 Logical Indexing

Another way to speed up code and avoid loops is to use *logical indexing*, which is described below in a pair of examples. Logical indexing can be very useful when we want to modify some of the elements in an array depending on whether they meet a particular condition.

Before looking at the examples, we show one way to generate data that we can use. The built-in `magic` function returns a square array of numbers with the 'magic' property that the rows, columns and diagonals all add up to the same value. The command

```
>> A = magic(6)
ans =
    35     1     6    26    19    24
     3    32     7    21    23    25
    31     9     2    22    27    20
     8    28    33    17    10    15
    30     5    34    12    14    16
     4    36    29    13    18    11
```

produces an array `A` representing a magic square which contains all the numbers from 1 to 36 (6 × 6) inclusive. You can check whether the column, row and diagonal sums are equal.

The next example illustrates code that does *not* use logical indexing. This is then followed by an example that does.

■ Example 9.8 (No logical indexing)

O9.A

Here, we create a new array `B` with the same size as `A`. `B` has a value of one everywhere unless the corresponding element in A is divisible by three, in which case the value is set to zero. We can do this by looping over all the elements in A, testing each one in turn and assigning the value to B:

```
N = 6;
% NxN magic square
A = magic(N);
% Initialise B to be the same size as A with ones
B = ones(N);

% Set B to one where A is divisible by 3
% Loop over all elements
for row = 1:N
  for col = 1:N
    % Get current element's value.
    value = A(row,col);
    % Is it a multiple of 3?
    if ( mod(value, 3) == 0 )
      B(row, col) = 0;
    end
  end
end

disp(B)
```

■

■ Activity 9.6

Download the code for Example 9.8 from the website and run it. Confirm that this code produces the array B as shown below. Note the pattern in the values of B, and confirm that it matches what you expect based on the values in the array A.

O9.A

```
     1     1     0     1     1     0
     0     1     1     0     1     1
     1     0     1     1     0     1
     1     1     0     1     1     0
     0     1     1     0     1     1
     1     0     1     1     0     1
```

Note how the code loops over all the elements, testing each one separately to see if it is divisible by 3. This is done with the test

```
if ( mod(value, 3) == 0 ) ...
```

The mod function returns the remainder when value is divided by 3. ■

We can obtain a speed-up in the above code by using logical indexing, a vectorized operation that allows loops to be discarded.

■ Example 9.9 (Using logical indexing)

O9.A

Example 9.8 can be carried out with logical indexing as follows:

```
N = 6;
A = magic(N);
B = ones(N);

indices = mod(A, 3) == 0;
B(indices) = 0;

disp(B)
```

The original pair of nested loops have been replaced by two lines. There is quite a lot happening in these lines, so let's look at them in more detail. The call `mod(A,3)` applies the function `mod` *element-wise* to `A` to calculate the remainder for every element after dividing by three.

Next, an equality test (`mod(A,3)==0`) compares every element in the array of remainders with zero. This is again an element-wise step and it results in an array of *logical* values: `true` values indicate where the remainder was zero, otherwise the values are `false`.

The array of logical values (with `true` indicating the elements of interest) is assigned to a variable `indices`. This is a logical indexing variable and it is used in the next line. It is applied directly to the array `B` with the round (indexing) brackets. Logical indexing causes the assignment to 0 to be carried out for all elements in the array `B` for which the corresponding element in the variable `indices` is `true`.

These two lines of code can even be replaced with a single line by avoiding the use of the variable `indices`:

```
B(mod(A, 3) == 0) = 0;
```

■

■ Activity 9.7

O9.A

Modify the code in Example 9.9 so that it starts with a magic square of size 5×5 and produces a second array that has a value of zero where the magic square is odd and a value of 10 where the magic square is even. ■

Example 9.9 uses much more dense and compact code and is much faster than the original longer code which used of a pair of nested loops. The difference between them illustrates an important point:

Code that has been optimized for efficiency tends to be harder to read.

Generally, it is better to start with code that might be slow and inefficient, but where it is clear what it is meant to do. Optimizing the code should only be done (a) if it is considered important and (b) once we are sure that the code is running correctly. This illustrates a second key point:

It is more important that code runs correctly than that it runs efficiently.

These points may seem obvious but it is fairly common for coders to forget them during routine programming.

9.3.4 A Few More Tips for Efficient Code

These tips are adapted from the MATLAB help sections:

- Keep *m*-files small. If a file gets large, then break it up and call code in one file from another.
- Convert script files into functions whenever possible. Functions generally run a little faster than scripts. See Sections 3.2 and 3.4 for more information on function and script files.
- Keep functions themselves as small and as simple as possible. If a function gets very long, look for parts of it that can be put into a new function and call it from the first one. (Recall that the second function can still be in the same file as the first.)
- Simple utility functions that are called from many other functions can be put into their own *m*-files so that they can be accessed easily (their location will need to be in the MATLAB path; see Section 3.7).
- *To repeat*: Avoid loops where possible. Try to use vectorization and element-wise operations.
- *To repeat*: Pre-allocate arrays when it is known how large they will be. Try to avoid growing a loop with each iteration of an array if possible, especially if there are likely to be many iterations. If the size of an array is not known in advance, it is sometimes possible to pre-allocate it to a size that is safely larger than what will be needed.

■ Activity 9.8

O9.A, O9.C

In Exercise 3.9 you wrote code to perform respiratory gating using a bellows signal. Make a copy of your solution to Exercise 3.9 (or the one from the book's web site) and adapt it so that the gating operation is implemented in two different ways: using a loop and using logical indexing. Add commands to measure the time taken to run each implementation.

Note that the code to compute the low and high bounds of the gating window should remain unchanged – you only need to modify the code for the gating operation itself (i.e. computing the indices of all elements in the full bellows array that are within the gating window).

Of the two implementations, which is the most time efficient and which do you think is the most easy to understand? ■

9.4 RECURSIVE AND DYNAMIC PROGRAMMING

The way some problems are defined can lead to time consuming or memory demanding solutions when a program is written to solve them. A recursively defined problem is one such type of problem. When writing a program to solve a recursive problem it can be easy to write a solution that naively ends up having a high computational cost. Dynamic programming is one way to try and reduce this cost. Before discussing this in more detail, we will give a recap of recursive programs focusing particularly on the number of function calls that are made.

The following example is a continuation of Activity 3.11 that looked at the factorial function. Here, we will concentrate on the number of times that the function is called and with what arguments.

■ Example 9.10

O9.D

The code for evaluating the factorial $n!$ of a number n is repeated here but we have changed the name slightly so that, for brevity, it is $f(n)$.

```
function result = f(n)
% f(n) : return the factorial of n

if (n<=0)
  % Stopping condition.
  result = 1;
else
  result = n * f(n-1);
end

end
```

If we want to evaluate f for $n = 5$ say, then the call we make will be $f(5)$ and this will in turn make a call to evaluate $f(4)$ which in turn makes a call to evaluate $f(3)$ and so on. The sequence of calls can be written

$$f(5) \rightarrow f(4) \rightarrow f(3) \rightarrow f(2) \rightarrow f(1) \rightarrow f(0)$$

The final call to $f(0)$ is dealt with by the *stopping condition* (see Section 3.10) in the `if` clause and no further recursive calls are made. In other words, after the original call to $f(5)$ a further five calls to the function are made recursively before stopping. In the general case, to evaluate $n!$, we will carry out n recursive calls of the function f. ■

This type of analysis of recursive function calls can be performed for many recursive implementations. In the case of the factorial function above the number of recursive calls was not excessive, and so the implementation shown was not particularly inefficient. Now let's consider another example which *does* turn out to have an inefficient solution if implemented naively.

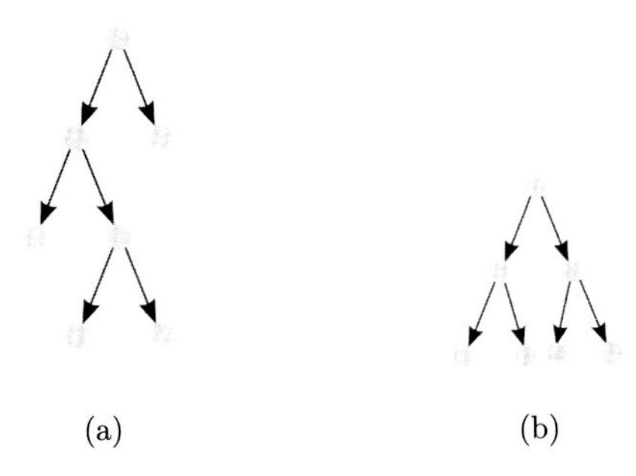

(a) (b)

FIGURE 9.1 Examples of full binary trees with seven nodes.

Full Binary Trees: We define a *full* binary tree as one with a set of nodes of which one is a *root* node and every node in the tree is either a *leaf* node or is the *parent* of exactly two *child* nodes. A leaf node is one with no children. In the example shown in Fig. 9.1a, there are seven nodes shown as gray circles. Each arrow connects a parent node to a child node. The root node is at the top of the tree, and there are four leaf nodes and two intermediate nodes (neither a root nor a leaf).

A natural question to ask is *how many different full binary trees exist for a given number of leaves*? The above example is one with four leaf nodes. How can we calculate how many such trees are possible with four leaf nodes?

One way to answer this question is to look at *sub-trees* below the root node. In the above example, the sub-tree below the root node on the left contains three leaf nodes and the sub-tree on the right contains the remaining single leaf node (i.e. the right-hand sub-tree consists of one node only).

In another tree with four leaf nodes, we could for example have two leaves in the left sub-tree and two leaves in the right sub-tree. An example is shown in Fig. 9.1b.

Considering the sub-trees on either side of the root node provides us with a way to find the total number of trees for a given number of nodes. For the four leaf node example, we can have the following pairs of numbers of leaves assigned to each sub-tree beneath the root node:

$(1, 3) \quad (2, 2) \quad (3, 1)$

Let T_k be the number of trees with k leaf nodes. For each pairing above, we can work out the number of trees with that pairing by multiplying the number of possible sub-trees on either side of the root. For example, if we have a $(3, 1)$ split for the leaf nodes, then the total number of trees with this split will equal

$\#$ trees with three leaves $\times$ $\#$ trees with one leaf $= T_3\, T_1$

We can therefore obtain the total number of full binary trees with four leaves by adding up the expressions over all possible splits:

$T_4 = T_1 T_3 + T_2 T_2 + T_3 T_1$

By writing the formula for T_4 in this way, we can start to see the recursive nature of a function that would work for *any* number of nodes. The value of T_4 above depends on the previous values of T_k for $k = 1, 2, 3$. We can write a similar expression for five leaves, six leaves, etc.:

$T_5 = T_1 T_4 + T_2 T_3 + T_3 T_2 + T_4 T_1$

$T_6 = T_1 T_5 + T_2 T_4 + T_3 T_3 + T_4 T_2 + T_5 T_1$

...

A general formula for n leaves can be written as

$$T_N = \sum_{k=1}^{N-1} T_k \, T_{N-k}$$

We haven't actually calculated any values of T_k yet but note that if we have only a single node, i.e. $k = 1$, the root and the leaf are the same node. So there is only one possible tree which means that $T_1 = 1$. This can be used for the stopping condition (ground case) of our recursive definition.

■ Example 9.11 (A naive and inefficient recursive program to determine the number of trees)

O9.D

We can see from the definition of T_N above that it is recursive and this leads to an obvious possible recursive program to evaluate it:

```
function nTrees = T(N)
% Stopping condition:
if N == 1
  nTrees = 1;
  return
end

% N > 1 here.
% Initialise value to calculate.
nTrees = 0;

% The main loop.
for k = 1:N-1
  nTrees = nTrees + T(k) * T(N-k);
end

end
```

The code has a stopping condition for $N = 1$, in which case there is a single leaf node. This would correspond to a single node in the whole tree and

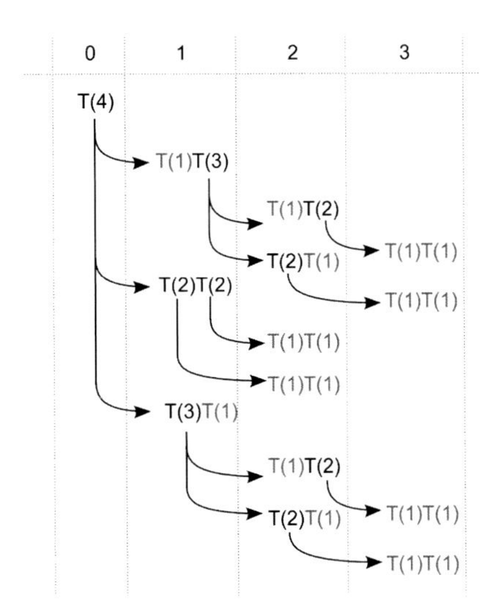

FIGURE 9.2 A visualization of the calls made when counting the number of full binary trees with 4 nodes.

so there is only one possible tree. For larger values of N, recursive calls are made for smaller values (k and $N-k$) and the results are multiplied together and accumulated. *N.B.* This code has been written compactly and without usage or many comments to save space. This is not an example of perfect programming. ■

In the above example, consider the case $N = 4$. We can write out in full the calls that are made:

$T(4) = T(1)T(3) + T(2)T(2) + T(3)T(1)$

We see that T is called twice with the stopping condition $(N = 1)$, twice with $N = 2$ and twice with $N = 3$. Note that these calls will lead to further calls:

- Each call with $N = 3$ leads to calls $T(1)T(2) + T(2)T(1)$, i.e. two further calls with $N = 1$ and $N = 2$ each.
- Each call with $N = 2$ leads to calls $T(1)T(1)$.
- Each call with $N = 1$ returns the stopping condition value and ends the recursion.

A visualization of the calls made when evaluating $T(4)$ with this function is given in Fig. 9.2. The numbers along the top indicate the 'depth' of the recursion for each call. Recursion depth 0 corresponds to the original (first) call which, in turn, makes calls to evaluate three pairs of terms at recursion depth 1 to find their products and add the results. These make calls at recursion depth 2 and so on. The calls when $N = 1$ are marked in red as they match the stopping condition, at which each sequence of recursive calls is ended.

Adding up all these calls, including those within recursive ones, we can find the total number of calls. There is 1 call with $N = 4$, 2 calls with $N = 3$, 6 calls with $N = 2$ and 18 calls with $N = 1$. In other words, there are many repeated calls for smaller values of N that are really unnecessary because they are repeatedly calculating the number of trees for the same value of N. This shows that the naive recursive implementation can lead to a very inefficient solution in which a lot of work is duplicated.

9.4.1 A Note on the Depth of Recursive Function Calls

We saw in the previous example that the call to evaluate $T(4)$ led to recursive calls to a depth of three. In general, for this example, a call to evaluate $T(N)$ will lead to a depth of $N - 1$ in the final level of recursion. There are practical limits when running recursive functions and MATLAB has a built-in limit that stops the levels of recursion exceeding a certain number (typically 500). This is sensible because each call uses up a certain amount of memory and if too many calls are made, then this would use up all of the computer's memory. Also, it is possible for a programmer to write a recursive function that, perhaps due to a bug, fails to hit its stopping condition. In this case we should guard against the function recursively calling itself an infinite number of times. In general, we should take care when using recursive functions and avoid using them when the depth of recursion is likely to be high.

■ Activity 9.9

O9.D

Pharmacokinetics is a branch of pharmacology that studies what happens to chemicals (such as drugs) administered to the body. When doctors administer a drug to a patient, the kidneys act to filter it out of the blood stream, so the level of the drug in the body will continually drop unless extra doses are given. When administering regular doses of a drug, doctors are interested in modeling what the level of the drug will be in the body over time. This activity addresses this issue.

We first make a simplifying assumption: each day, the kidneys will filter out a fixed proportion of the drug that is still in the body. This assumption leads to an intuitive recursive definition of the problem of modeling drug levels in the body:

$$\text{level}_n = \text{decay} \times \text{level}_{n-1} + \text{dose}$$

That is, the level of drug at day n is equal to the level at day $n - 1$ multiplied by a factor by which the level decays each day (due to filtering by the kidneys) plus the new dose administered each day. We assume that the level at the first day is zero (i.e. this is the stopping condition).

Write a recursive MATLAB function *m*-file to compute the level of drug in the body at day n, given a decay factor and a daily dose as additional arguments. Write a script *m*-file that reads in the dose, decay factor and a maximum number of days from the command window, and then uses the function to determine the drug level for all days up to and including the maximum day. The script should then produce a plot of drug level against day.

Analyze your solution to determine the maximum depth of recursive call and the number of recursive calls that will be made for a given value of the day n.

Also write an iterative version of the function and compare the time efficiency of the two implementations. Which is the most time efficient and which is the easiest to understand? ■

9.5 DYNAMIC PROGRAMMING TO IMPROVE PERFORMANCE

The previous example that counted full binary trees is one that can benefit from *dynamic programming*. This is because it is computed by evaluating the same function on sub-problems of the original problem (in our case on sub-trees of the original tree).

The basic idea of dynamic programming is to store results that might need to be repeatedly calculated. This is known as *memoization*. We keep a 'memo' of the results that have already been calculated and that may be useful later. Values of the function are only directly evaluated if they have not previously been stored (memoized).

■ Example 9.12 (Using dynamic programming for the tree counting example)

O9.D, O9.E

In order to reduce the number of evaluations on the above tree counting example, we will write the code differently using a dynamic programming approach. We begin by writing the top-level function. This initializes an array for the memoized values and calls a helper function called `count`:

```
function nTrees = countTreesDynamic(N)

memoValues = -1 * ones(1,N);
nTrees = count(N, memoValues);

end
```

The helper function accepts two arguments, the value of N for which the tree count is required and the array of stored values. The stored value array (`memoVals`) is initialized so that every entry is -1. We use a value of -1 to indicate that the result has not yet been calculated. At the start, this is the case for all of the possible values of N. Now we can consider the helper function. The start of this function looks like this:

```
function [nTrees, memoVals] = count(N, memoVals)
% Helper function for counting the trees. Keeps
% a memoised list of values to prevent too many
% calculations.

% N > 1. Check if we have already
% calculated for this value of N.
if memoVals(N) > -1
  nTrees = memoVals(N);
  return
end
% ... continues below ...
```

Note how the function accepts the number of leaves N and the array for the stored values and returns two items: the number of trees and the array of stored values. The array of stored values also appears as a return value because it may be updated during the call. The function begins by seeing if the required value has already been calculated. This is done by checking if the value stored at index N in `memoVals` array is greater than -1 (the initialization value of each entry). If it is greater than -1, then the value was calculated earlier and we can assign it directly to the output argument `nTrees` and return. Job done. There is no need to update the `memoVals` array.

Now let's look at the next part of the function:

```
% ...
% stopping condition:
if N == 1
  nTrees = 1;
  memoVals(N) = nTrees;
  return
end
% ...
```

This part deals with the stopping condition when there is only one leaf in the tree. In this case there is only one possible tree and this is the count returned. Importantly, this count is stored in the `memoVals` array (at index 1) so that we can use it in any later repeated calls with $N = 1$.

The remaining parts of the function are only reached if $N > 1$ and the number of trees for N has not yet been calculated and memoized.

```
% ...
% No recorded value. Need to calculate from scratch.
% Initialise:
nTrees = 0;

for k=1:N-1
  [nLeft , memoVals] = count(k, memoVals);
```

```
    [nRight, memoVals] = count(N-k, memoVals);
    nTrees = nTrees + nLeft * nRight;
end

% Store value we have just calculated so we don't
% need to do this again.
memoVals(N) = nTrees;

end
```

Here, we make two recursive calls to the same function. Each time we collect both arguments, the count of the trees (`nLeft`/`nRight`) and also the (possibly updated) `memoVals` array. During the loop, the number of trees is calculated and after it is complete, we carry out the important step of storing the result in the array at the index N. ■

In the exercises, we will compare the speed of this memoizing version of the tree counting function with the speed of the naive recursive implementation. It is sufficient to say here that the memoized version is much faster because it avoids a lot of unnecessary duplication of calculations of values which have previously been calculated and stored.

We note that the use of a memoized array of previously calculated and cached results has led to an increase in memory use. In the above case, this has been a small price to pay compared to the gain in speed but, in general, we should bear in mind the memory used when seeking to adopt this approach to ensure that the memory usage of the program remains reasonable.

9.6 SUMMARY

This chapter has covered some aspects of code efficiency and general tips to help improve it. These form useful advice but it is important to maintain good coding practice (see Chapter 4). The general advice when coding is to proceed in two broad stages:

- Write code that is correct, clear, readable and tested.
- Where possible, only consider performance and efficiency when the above step is complete. In particular:
  - Decide whether or not improvements in efficiency are necessary.
  - Only implement them if they are, by rewriting appropriate parts of the code.

The reason for this is that code that has been optimized to be as efficient as possible tends to be less readable and harder to debug when errors arise. Code that takes up more space and uses more operations might be less efficient but is more likely to be clearer and easier to follow.

To stress this point: it is more important to have code that is *correct* than code that is fast or memory efficient. Once we are confident that the code is correct, we can consider changes for optimizing efficiency.

In other words, it is almost always okay to start off with code that works and is slow. We can then incrementally change it so that it becomes quicker, only changing small parts of the code at a time (in line with the incremental development model outlined in Section 4.2).

In the second half of the chapter we looked specifically at recursive functions and how, if implemented naively, they can be inefficient by carrying out a lot of unnecessary calculation. We took an idea from dynamic programming, in which we store or cache the results of calculations, to avoid such duplication. Generally this will lead to significant improvements in the calculation of recursive functions but we should remain aware that it can cause an increase in memory usage.

9.7 FURTHER RESOURCES

- The MATLAB built-in help has a number of tips on improving the performance of programs from a time or memory point of view. A good place to look is under *MATLAB → Advanced Software Development → Performance and Memory → Code Performance*. This contains several sections, for example it has a section on vectorization.
- Dynamic programming applies to a number of other areas:
  - For a good general description of how it applies to a number of interesting problems follow the links from: http://people.cs.clemson.edu/~bcdean/dp_practice/.
  - A more general tutorial can be found here: http://www.codechef.com/wiki/tutorial-dynamic-programming.
  - A good list of challenge problems can be found at: http://uva.onlinejudge.org/index.php?option=com_onlinejudge&Itemid=8&category=114.
- Notes on memoization are available at: cs.wellesley.edu/~cs231/fall01/memoization.pdf.
- The number of trees with a given number of leaves is an example of an application of Catalan numbers which are interesting in their own right and have a diverse number of other applications: http://en.wikipedia.org/wiki/Catalan_number.

EXERCISES

■ Exercise 9.1

O9.B

a. Calculate how much memory will be required by the following assignment

```
x = ones(1000)
```

b. Explain why a call to `z = 3 * zeros(1000000)` is likely to cause problems when it is run. ■

■ Exercise 9.2

O9.A
O9.B

a. Give two reasons why the following code is not efficient

```
myArr = 0;
N = 100000;

for n = 2:N
   myArr(n) = 2 + myArr(n-1);
end
```

b. Re-write the above code so that it runs more efficiently.
c. Use the built-in timer functions to estimate the speed-up given when the above code is re-written. ■

■ Exercise 9.3

O9.A

Re-write the following code so that it becomes more time efficient.

```
N = 20;

a = [];
for j = 1:N
  a(j) = j * pi / N;
end

for k = 1:numel(a)
  b(k) = sin(a(k));
end

for m = 1:numel(a)
  c(m) = b(m) / a(m);
end

plot(a, b, 'r')
hold on
plot(a, c, 'k')
legend('sin', 'sinc')
```

■

■ Exercise 9.4

O9.A

a. Use the built-in random function `rand` to generate a 5×6 array of uniformly distributed random numbers, and assign it to a variable called `A`.
b. Use loop(s) to check the elements in `A`. If an element is less than 0.6, replace it with a value of 0. Do this inefficiently using the same approach as in Example 9.8.

c. Use logical indexing to achieve the same result as in the previous part, i.e. by avoiding loops.
d. Use logical indexing to set the value of any elements in `A` that are greater than 0.7 to a new value of -1.
e. Use logical indexing to set any values in `A` that are greater than 0.7 *OR* less than 0.3 to -1.
(*Hint: Use the bitwise logical or operator:* |). ■

■ Exercise 9.5

O9.A, O9.B, O9.C

a. Write a MATLAB function that takes a single integer input argument N and computes a 2-D array representing the multiplication table for integers from 1 up to N. For example, using the `disp` command to display the table for $N = 5$ should show the following array:

```
1  2  3  4  5
2  4  6  8 10
3  6  9 12 15
4  8 12 16 20
5 10 15 20 25
```

Use loops in your implementation and focus on writing clear, understandable code that works.
b. Now write another implementation, this time optimized for speed. Run and time both implementations for values of N from 1 to 200 and show the time results for both methods.
(*Hints: You can use matrix multiplication or the* `repmat` *command to obtain faster implementations. The commands* `tic` *and* `toc` *can be used for timing.*)
c. Write code to run both functions for all values of N from 1 to 200 calculate the total time taken for each one. Write code to display the time results for both methods.
(*Hint: The commands* `tic` *and* `toc` *can be used for timing.*)
d. Finally, write a function that just *displays* a multiplication table for a given size and that uses very little memory (i.e. one that is memory efficient). N.B. It does not need to store the table. ■

■ Exercise 9.6

O9.A, O9.C

In Exercise 5.7 you wrote a MATLAB script *m*-file to determine the recommended drug dose based on the results of a phase I clinical trial. This involved searching through a series of toxicity results for a number of cohorts of patients (with three in each cohort) to find the first instance in which two or more of the three patients experienced significant toxicity.

Make a copy of your solution to Exercise 5.7 (or the one from the book's web site) and adapt it so that this searching is performed in two different ways. One should use a loop and the other should use logical indexing. It should be possible to write the version that uses logical indexing with a single line of code. Make sure that both implementations give exactly the same results when applied to the data in the file *phaseI_data.mat*. Add commands to your script *m*-file to measure the time taken to run each of the two implementations.

Leave the remainder of the code unchanged, i.e. the code that finds the recommended dose once the first cohort with two or more significant toxicity results has been found.

Which implementation is the most time efficient? Which is the easiest to understand? ■

■ Exercise 9.7

O9.A, O9.C

In Exercise 3.6 you wrote a function to determine the indices of all peaks and troughs in a given array, and demonstrated its use for gait analysis. Make a copy of your solution to Exercise 3.6 (or the one from the book's web site) and adapt it so that there are two different functions that each identify the peaks and troughs, but in different ways. One should use a loop and the other should use logical indexing. Make sure that both give exactly the same results when applied to analyzing the z-coordinates of the ankle marker as in Exercise 3.6 (i.e. the file *LAnkle_tracking.mat*). Add commands to your main script *m*-file to call and measure the time taken to run each of the two functions.

Which of the two implementations is the most time efficient? Which is the easiest to understand? ■

■ Exercise 9.8

O9.A, O9.D, O9.E

Go to the book's web site and download both versions of the code that counts the number of full binary trees with N leaves:

- The naive recursive version (Example 9.11).
- The version that uses memoization (Example 9.12).

a. Write a MATLAB script to measure the execution times of the two implementations for a single value of N. Be careful when choosing the value of N. When using the naive version, a value of N greater than around 14 could lead to a long wait!
b. Extend your script to compute the execution times for the two implementations for $N = 1 \ldots 10$, and plot both sets of values on the same graph against N. What do you notice about the times taken by each method as N increases? ■

■ Exercise 9.9

O9.A
O9.D
O9.E

The Fibonacci sequence is the series of integers $1, 1, 2, 3, 5, 8, \ldots$ and it is defined mathematically and recursively by the formula

$$f_1 = 1, \qquad f_2 = 1, \qquad f_{n+1} = f_n + f_{n-1}, \quad \text{for } n > 2$$

Recall that Exercise 3.10 involved writing a recursive implementation of a function to compute Fibonacci numbers. The Fibonacci sequence has applications in a wide range of fields including biology [4].

a. Write a function that directly implements the recursive Fibonacci formula. Name the function *fibonacciRecursive* and save it to a function *m*-file.
b. Each call to the function will make further calls recursively with smaller values of n. If the function is called originally (recursion depth 0) with an input argument of $n = 5$, work out how many calls there are for the function with arguments of $n = 4, 3, 2$ and 1, and how many calls in total.
c. Extend this to work out the number of all such calls to the function when we start with higher values of n, i.e. $n = 6$, $n = 7$, etc.
d. One way to avoid the overhead of multiple calls which evaluate the same value of the function in Exercise 9.9 is to avoid recursive calls altogether. One way to do this is to calculate the values in a 'bottom up' way, starting with the values for the lowest values of n and 'building up'. This way, we can visit each value of n once only.
   By defining values for the current and previous terms in the sequence (or otherwise) write a function that loops upward to the required value of n. Within each iteration, the following will need to be done:
   - Calculate the value of the next term in the sequence.
   - Increment a counter that keeps track of how far we have moved along the sequence.
   - Update variables holding the previous and current terms based on the incremented counter.
     (Hint: Use the value calculated in the first step).
   - Check if the loop has reached the required value of n, stopping if this is the case and returning the required value.

   N.B. Such *bottom-up evaluation* is another technique used in dynamic programming (as well as memoization).
e. Estimate the speed up of the dynamic programming bottom-up evaluation of the Fibonacci code compared with the previous recursive version. How many times faster is the new code? It may be advisable not to call the previous version with a value much higher than 30 or so. ■

FAMOUS COMPUTER PROGRAMMER: RICHARD STALLMAN

Richard Stallman is an American software freedom activist and computer programmer, who was born in 1953 in New York, USA. Whilst at high school, he got a summer job at the IBM New York Scientific Center.

He was tasked with writing a numerical analysis program in FORTRAN, an early programming language. He completed this software within 2 weeks and spent the rest of the summer developing a text editor for his own enjoyment. Later, he studied Physics at Harvard, and also worked as a programmer at the Massachusetts Institute of Technology (MIT).

Whilst at MIT he objected to the introduction of passwords to limit access to the laboratory's computers. He wrote a computer program to access and automatically decrypt all users' passwords and sent them messages containing their decrypted password, along with a suggestion to change it to an empty password, i.e. to reinstate free access to computers. Although some users followed his suggestion, eventually password control became the norm in computer access.

Stallman is best known as the originator of the GNU software project (www.gnu.org), which aims to promote the concept of free software. Stallman started the GNU project in 1983, and it has gone on to become one of the major influences on software production and distribution in the world today. Stallman has not benefited financially from his important work on the GNU project and professes to care little for material wealth.

Stallman has been a passionate and outspoken advocate for software freedom, and his outspokenness has made him enemies as well as friends. He has described Steve Jobs, the Apple co-founder, as an "evil genius" for building "computers that were prisons/jails for the user and to make them stylish and chic". Linus Torvalds (see Chapter 11's Famous Computer Programmer), the developer of the LINUX open-source operating system, has criticized Stallman for what he considers "black-and-white thinking" and bringing more harm than good to the free software community. Today, Richard Stallman lives in Cambridge, Massachusetts, and devotes his time to free software advocacy and the GNU project.

> *"If programmers deserve to be rewarded for creating innovative programs, by the same token they deserve to be punished if they restrict the use of these programs."*
>
> **Richard Stallman**

CHAPTER 10

Signal and Image Processing

LEARNING OBJECTIVES

At the end of this chapter you should be able to:

O10.A Load, save, and visualize signal and image data with MATLAB commands or, in the case of image data, with the available interactive tools

O10.B Carry out some of the fundamental 1-D signal processing steps with MATLAB

O10.C Carry out some of the fundamental image processing steps with MATLAB

10.1 INTRODUCTION

Signals and images form an important part of general data analysis and have a particularly important role to play in many biomedical applications. We can carry out a number of tasks involving them: collecting the raw data, storing and retrieving data from files, visualization, processing and so on.

It is very common to collect biomedical signals that vary in time. These can be generated, for example, by monitoring physiological processes in the human body. We can measure the heart rate, blood glucose levels, electrical activity in the brain and so on. Each of these can be represented by a time-varying signal which we can process and analyze to derive further information and devise biomarkers relating to the health of patients. Fig. 10.1 shows some examples of such time-varying signals.

An image can be viewed as a more general signal. For example, an image acquired with an everyday digital camera represents a set of measurements that are arranged in *space* on a 2-D grid of pixel locations. This bears much similarity with signals acquired on a 1-D regular 'grid' of *time* points.

Images are an important source of information in biomedical applications as they can assist clinicians in making a diagnosis. Biomedical examples of 2-D images include microscopy images of cells and photographs of skin for de-

MATLAB Programming for Biomedical Engineers and Scientists. DOI: 10.1016/B978-0-12-812203-7.00010-0

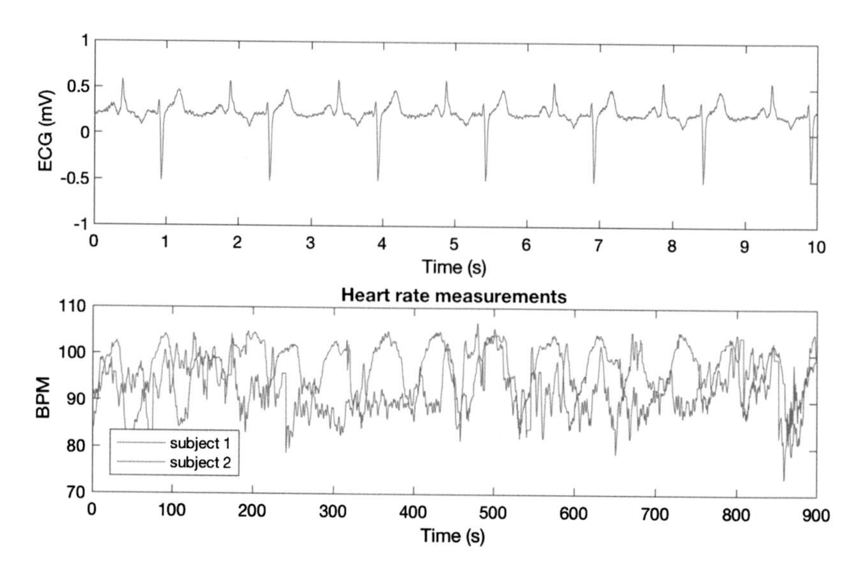

FIGURE 10.1 Top: ECG measurements for a 10 second interval. Bottom: Heart rate measurements over a 15 minute interval for two different subjects.

tecting melanomas. Magnetic resonance (MR) scans of organs are examples of 3-D biomedical images. Positron emission tomography (PET) scanners can dynamically acquire multiple 3-D images of an organ over time to show how its metabolic function varies. This is an example of 4-D medical image data (i.e. three spatial dimensions and one time dimension). Fig. 10.2 shows some examples of different types of medical image.

Processing images is an important part of developing biomedical applications and is typically carried out using computer programming. Specialist software as well as general programs such as MATLAB can carry out a variety of tasks that manipulate and alter images in ways that can provide useful information. This chapter will focus on some of the basic steps of signal and image manipulation and processing. A few methods will be described in general terms and we will also show how they can be carried out in MATLAB.

10.2 STORING AND READING 1-D SIGNALS

1-D signals, such as time series data, have a very simple structure and can be easily stored in text files with, say, one measurement on each line of the file. Typically, we also need 'meta-data' to indicate what units are used for the measurements. We also need to know the sampling frequency, i.e. the time step between adjacent measurements. If the measurements are not obtained at regularly spaced time points, then we can store the times at which each measurement was taken as well as the value of the measurement. This remains

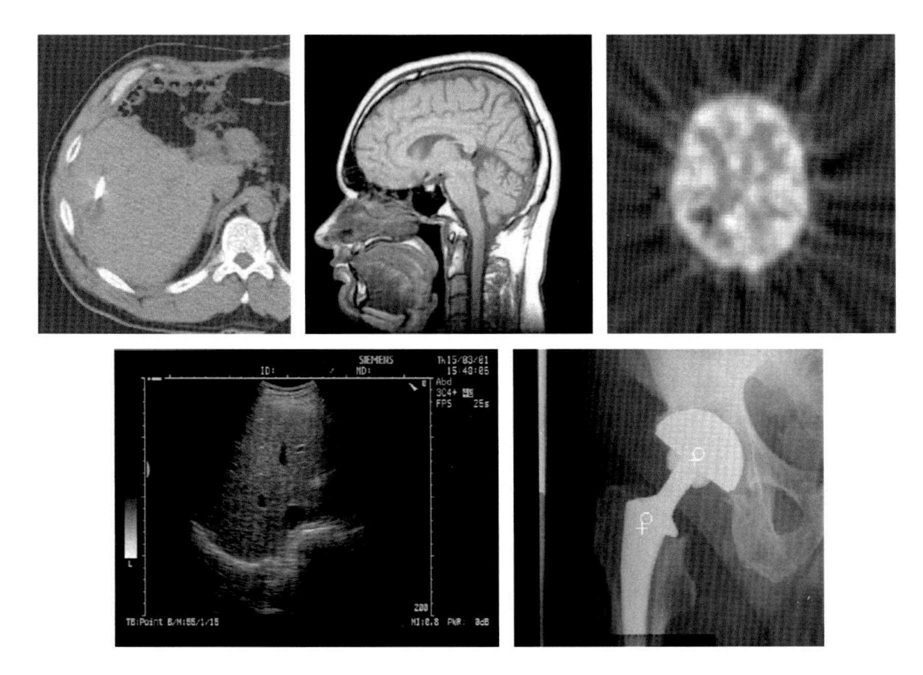

FIGURE 10.2 Examples of medical images: from top-left, slices taken from a computed tomography (CT) image of the abdomen, an MR image of the head, a single PET volume of the head, a 2-D ultrasound image of the abdomen and a 2-D X-ray of a hip.

a relatively simple structure and can, for example, be stored in a 'comma-separated values' (*.csv*), file with two columns, one for the times and one for the measurements. Of course it is also possible to store data for a 1-D signal in a binary file format. The different methods for saving and reading both text and binary format data files were discussed in Chapter 6.

10.3 PROCESSING 1-D SIGNALS

We have in fact already seen a number of different examples in the book in which we analyzed and extracted information from 1-D signals.

For example, in Activity 1.8, Exercise 2.8, and Exercise 3.6, we looked at data to represent the gait of a person when walking. In particular, we treated the z-coordinate of the ankle (its height) as a time-varying 1-D signal and searched for peaks and troughs in the data. In Activity 5.8, we also visualized this height data. In Exercise 3.8, we looked at heart rate data to find times at which there were abnormal changes. We looked at time series data from an oscillatory monitoring device in Exercise 4.4 and examined how we could use it to obtain blood pressure and pulse rate information.

Other examples include Exercise 3.9, which looked at a 1-D signal from a respiratory bellows. This was further analyzed in Exercise 4.5 by smoothing the time-varying signal to reduce noise prior to finding peaks and estimating the length of the respiratory cycle. We considered data on blood sugar levels in Exercise 4.6 to identify time intervals where the average level indicates possible diabetes. In Example 4.1, we analyzed measurements of radial heart contraction (a time series for each of a number of different heart regions) to identify if a given patient exhibited dyssynchrony. In Exercise 7.2, Doppler ultrasound data, representing blood flow velocity, were used to estimate arterial resistivity.

These examples serve to illustrate that time series data in biomedical application is very common and the nature of the data and the processing applied to it can vary quite widely. In the next section, we consider in more detail a particular type of processing that is widely used to process both time series and imaging data.

10.4 CONVOLUTION

Convolution is an important and widely used tool in signal processing. It is a general method that can re-express many of the operations applied to signals, such as smoothing, finding peaks, etc. Convolution takes an array of signal values and a second array containing the values of a *kernel*. The kernel and the signal are *convolved* to produce a modified signal.

By way of an example, recall the signal from a respiratory bellows in Exercise 4.5. After reading in the raw data, the aim was to find peaks in the signal so that different breathing cycles could be distinguished. We assumed that a peak in the signal occurs when a signal value is greater than the values before it and after it (see Exercise 3.6). The problem is that some peaks detected in this way may be due to random noise and fluctuations in the data and may not be located at the boundaries of breathing cycles. This was why we *smoothed* the data so that the effect of noise was reduced and true peaks could be found more easily. Fig. 10.3 shows a plot of the raw bellows signal (red) and the smoothed data (blue). At $t = 4.6$ s, we can see an example of a 'false' peak in the original signal.

■ Example 10.1

O10.A, O10.B

The bellows plot used a signal sampled at 5 Hz, i.e. with a time resolution of 0.2 s. The code to produce the smoothed data is shown below.

```
% Load data
load('bellows_raw.mat');

% Sampling frequency
sampFreq = 5;
```

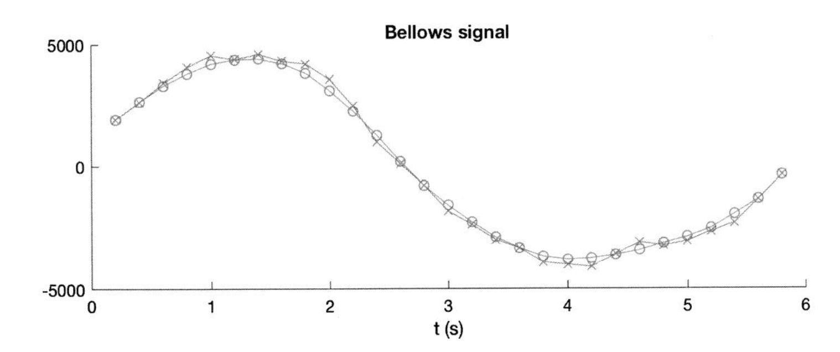

FIGURE 10.3 Data from a respiratory bellows measured over time: (red) original signal, (blue) smoothed signal.

```
% Compute times in seconds of measurements
t = (1:length(bellows_raw))/sampFreq;

% Smooth signal (See 'doc smooth')
bellows_smooth = smooth(bellows_raw);
```

■

The `smooth` function is part of the curve fitting toolbox in MATLAB and it works by calculating a moving average of values in a 'sliding window'. The window has a specific width and moves over the original signal. At each position, the average of the values it covers is calculated and assigned to the current value of the output signal.

By default, the `smooth` function uses a window width covering five values and this is illustrated in Fig. 10.4. Note in the figure how the mean value of a set of values is assigned at the central location, for example $(0.4 + 2.8 + 0.7 + 1.0 + 8.6)/5$ gives 2.7, $(2.8+0.7+1.0+8.6+6.9)/5$ gives 4.0 and so on. For a width 5 window, this means that the third value in the output sequence is the first one that has a full set of input values covered by the window. The first and second values are handled differently and how we handle them is a matter of choice. In the case of the `smooth` function, the first value in the output sequence is equal to the first of the input sequence and the second is the mean of the first three input values (rather than five). The values at the end of the sequence are treated similarly.

In general, when processing 1-D signal data in this way, it is common for values at (or near) the start and end, i.e. the boundaries, to be handled in a slightly different way from other non-boundary values.

In order to introduce the idea of convolution, note that finding the mean of a set of five adjacent values is equivalent to multiplying each by 0.2 and adding the results. We can picture a row of five weights, each equal to 0.2, being placed against the original sequence at successive positions. For each position, the

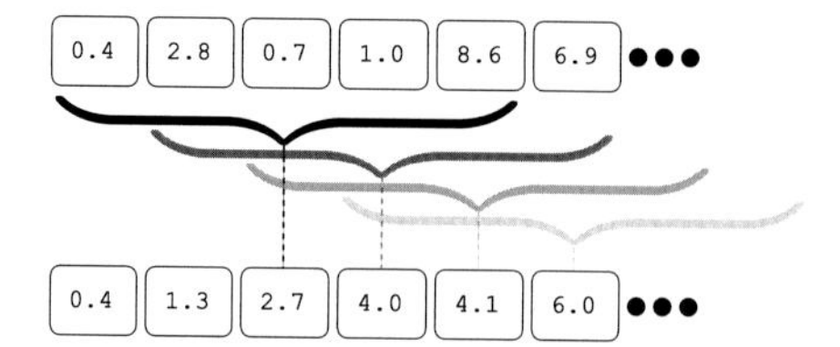

FIGURE 10.4 A sliding window is used to smooth a sequence of signal values. This illustrates a window of width five. Each successive set of five values are averaged to give the value in the output signal. The first few values of the input and output signals are shown. *N.B.* For this window, the third value in the output is the first one where the window 'covers' a full set of input values. The first two values need to be calculated differently (see text).

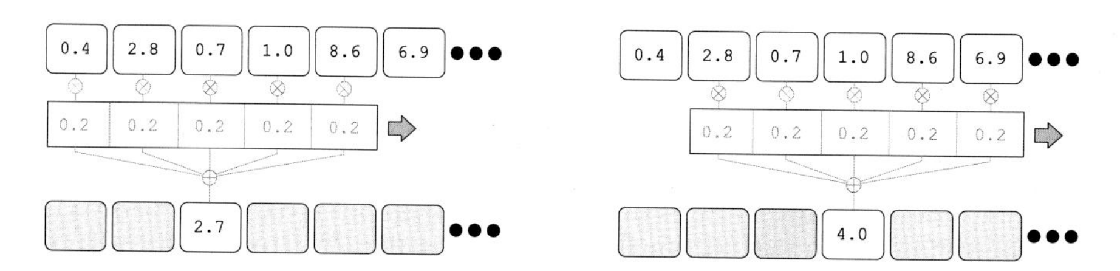

FIGURE 10.5 Smoothing a 1-D signal using a shifting row of weights. For each position of the weights, the original signal values are multiplied by the corresponding weights and the results summed.

values in the original sequence are multiplied by the corresponding weights and added together. After this step, the row of weights shifts forward one place and the process is repeated. This is illustrated in Fig. 10.5.

This is the fundamental idea of convolution. A set of weights in a *kernel* are multiplied by sets of adjacent values in the signal and the result is summed to give an output value. In simple terms, the *convolution of the signal with the kernel* is this 'multiply and add' process, applied after repeated shifts of the kernel.

■ Activity 10.1

O10.B

Now we can look at the process of smoothing using the MATLAB conv function. Type the following into a script and run it:

```
% The original 1-D signal.
u = [0.4, 2.8, 0.7, 1.0, 8.6, 6.9, 3.3, 10.2];
% The kernel.
kernel = [0.2, 0.2, 0.2, 0.2, 0.2];
% Do the convolution.
w = conv(u, kernel);
% What does the 'smooth' function give?
v = smooth(u);
```

Now we can compare the outputs from the `conv` and `smooth` functions. After executing the code above, the variable `w` should contain the following values.

```
0.08 0.64 0.78 0.98 2.7  4.0  4.1  6.0  5.8  4.08  2.7  2.04
```

The variable `v` contains

```
0.4  1.3  2.7  4.0  4.1  6.0  6.8  10.2
```

We can see immediately that there are differences between the two resulting outputs. They even have different lengths. This is partly due to differences in the way the boundaries are handled as we will see below. Before reading on, can you see which parts of the two outputs are the same? ■

In order to understand the output from `conv`, recall that the kernel has a length of five. When we slide the kernel across the array containing the original signal, we start with little overlap between the kernel and the signal. At the outset, there is an overlap of a single value, as shown below, where the top line shows the signal and the second line shows the kernel:

```
                    0.4  2.8  0.7  1.0  8.6  6.9  3.3 10.2
0.2  0.2  0.2  0.2  0.2
```

In this position, applying the kernel leads to the first element in the output array `w`, which is computed as the product of the first signal value and the rightmost kernel element: $0.08 = 0.4 \times 0.2$. The other kernel elements do not overlap the signal and are effectively multiplied by zero.

Applying successive shifts, the first four values of the output array `w` will still be obtained from a partial overlap of the kernel and the signal:

```
                                                          Result
                    0.4  2.8  0.7  1.0  8.6  6.9  3.3 10.2
0.2  0.2  0.2  0.2  0.2                                   -> 0.08

                    0.4  2.8  0.7  1.0  8.6  6.9  3.3 10.2
     0.2  0.2  0.2  0.2  0.2                              -> 0.64

                    0.4  2.8  0.7  1.0  8.6  6.9  3.3 10.2
          0.2  0.2  0.2  0.2  0.2                         -> 0.78

                    0.4  2.8  0.7  1.0  8.6  6.9  3.3 10.2
               0.2  0.2  0.2  0.2  0.2                    -> 0.98
```

■ Activity 10.2

Show the calculation that needs to be made to give the third value in the output (0.78). ■

O10.B

After computing the first four values, we can finally carry out convolution with the kernel fully overlapping the signal. This is shown in the following positions that give the next four values in w:

```
                                                Result
0.4  2.8  0.7  1.0  8.6  6.9  3.3 10.2
0.2  0.2  0.2  0.2  0.2                       -> 2.7

0.4  2.8  0.7  1.0  8.6  6.9  3.3 10.2
     0.2  0.2  0.2  0.2  0.2                  -> 4.0

0.4  2.8  0.7  1.0  8.6  6.9  3.3 10.2
          0.2  0.2  0.2  0.2  0.2             -> 4.1

0.4  2.8  0.7  1.0  8.6  6.9  3.3 10.2
               0.2  0.2  0.2  0.2  0.2        -> 6.0
```

Note how, for this set of values where the kernel fully overlaps the signal, the output from conv (variable w) matches the output from smooth (variable v).

Beyond this position, the remaining values in the convolution output will also result from partial overlaps of the kernel and will be calculated in the same way as the first four values above.

In summary, the outputs from conv and smooth have different lengths and the values at the boundaries are handled differently. It is reasonable to now ask: *How can we compare the outputs from the two different functions* smooth and conv?

■ Example 10.2

O10.B

In order to make the outputs comparable we can use an optional argument to the MATLAB conv function that specifies which part of the convolution output to calculate. We can achieve this with the following:

```
w2 = conv(u, kernel, 'same');
```

This option restricts the convolution to be calculated over a portion of the output that matches the length of the input signal (i.e. the variable u).

The resulting variable w2 now looks like this

```
0.78  0.98  2.7  4.0  4.1  6.0  5.8  4.08
```

which we can compare to the output from the smooth function (variable v from the earlier call) which we repeat here:

```
0.4   1.3   2.7  4.0  4.1  6.0  6.8  10.2
```

■

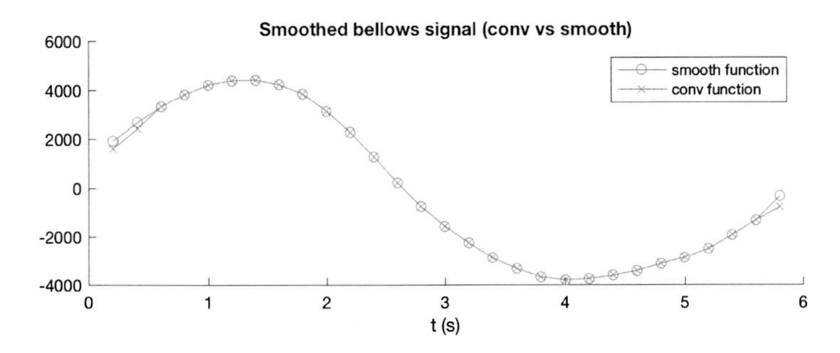

FIGURE 10.6 Smoothing the bellows signal using the MATLAB `smooth` and `conv` functions.

Here, we can see that the two outputs now match on the central portion (from the 3rd to the 6th elements, i.e. where the kernel has full overlap). The values at the ends still differ because they are handled differently by each function but this is not so significant in general, especially as the signals we typically work with will be much longer than the one given in this example and the majority of values in the array will not be near the boundary.

■ Example 10.3

We can repeat the smoothing of the bellows signal data in Example 10.1 using the `conv` function and compare the two results.

O10.B

```
% ... previous code as before
% Smooth signal (See 'doc smooth')
bellows_smooth = smooth(bellows_raw);

% Smooth using 'conv' function
kernel = [0.2, 0.2, 0.2, 0.2, 0.2];
bellows_smooth_2 = conv(bellows_raw, kernel, 'same');
```

■

If we plot the two resulting outputs we can see how they are the same at all values except for the sets of four values near each boundary (see Fig. 10.6).

10.4.1 Convolution: More Detail

So far, we have only considered a very simple kernel to illustrate the concept of convolution. It simply consisted of five equal values:

```
kernel = [0.2, 0.2, 0.2, 0.2, 0.2];
```

This has allowed us to ignore a further detail of how convolution is actually carried out. Specifically, when a kernel is convolved with an array, it is actually *reversed* before 'sliding' over the input signal values. As our kernel only contained a single repeated value, this reversal had no effect.

■ Activity 10.3

O10.B

We consider very simple example that illustrates this reversal of the kernel by using some simple integer arrays for our signal and kernel.

```
u = [1, 2, 1, 4, 3];
kernel = [1, 2, 3];
result = conv(u, kernel, 'same');
```

Displaying the variable `result` that contains the output of the convolution should now give

```
4    8    12    14    18
```

■

The calculations involved are illustrated below for each of the output values. In this case, the first and last values in the result are obtained with a partial overlap of the kernel. The important thing to note is that the kernel that is passed in `[1,2,3]` is reversed before being shifted across the input array:

```
   1  2  1  4  3
3  2  1               ->  4
---------------------------
   1  2  1  4  3
   3  2  1            ->  8
---------------------------
   1  2  1  4  3
      3  2  1         -> 12
---------------------------
   1  2  1  4  3
         3  2  1      -> 14
---------------------------
   1  2  1  4  3
            3  2  1   -> 18
---------------------------
```

For example, the third value is obtained by calculating $2 \times 3 + 1 \times 2 + 4 \times 1 = 12$.

> **The gradient of a signal:** Given a set of measurements in a 1-D signal and the time step between measurements, we can estimate the *gradient* at a point by using the *central difference*. If the nth value in a sequence is y_n and the time step between values is Δ_t, we can estimate the gradient g_n at that point using the values y_{n-1} and y_{n+1} and the formula
>
> $$g_n \approx \frac{y_{n+1} - y_{n-1}}{2\Delta_t}$$

■ Example 10.4

O10.A, O10.B

The file *pos-z.csv* is available on the book's web site and contains the

z-coordinates of a surgical robot arm with measurements taken every 400 ms (i.e. $\Delta_t = 0.4$ s). We will estimate the gradient of these z-values, i.e. their rate of change with respect to time. Before estimating the gradient, and as the data are very noisy, we aggressively smooth them using a wide kernel window (otherwise known as 'span') when applying the smoothing filter. We can load and smooth the data as follows:

```
zData = csvread('pos-z.csv');
% Aggressively smooth.
zSmooth = smooth(zData, 20);
```

Let the time-step variable be set by `deltaT = 0.4;`. We can now estimate the gradient of the signal at all points using convolution:

```
% A kernel to use with conv.
kernel = [-1, 0, 1] / (2 * deltaT);

% Flip the kernel before passing to conv
kernel = fliplr(kernel);

zGradient = conv(zSmooth, kernel, 'same');
```

Note how the kernel was flipped left to right before passing to `conv`. At the end points, not enough data are available to estimate the gradient (the kernel does not fully overlap the data). We can therefore arbitrarily set the values at the end points to zero.

```
zGradient(1) = 0, zGradient(end) = 0;
```

■

The full code for the above example is provided on the book's web site. In this case, the data represent z-positions so the gradient represents vertical velocity. If we plot the z-positions, the smoothed z-positions and the vertical velocity, for the data file specified, we obtain the plot shown in Fig. 10.7.

10.5 IMAGE DATA: STORING AND READING

Image data have a slightly more complicated structure than 1-D signal data such as time series signals. This means that we have take a little more consideration of what the image data represent. For example, images can contain a *lot* of data and this means we may need to be careful about what data type is used when storing an image. In general, we need to consider the following points:

- Discrete locations:
  - Image data are stored for a finite number of discrete locations.

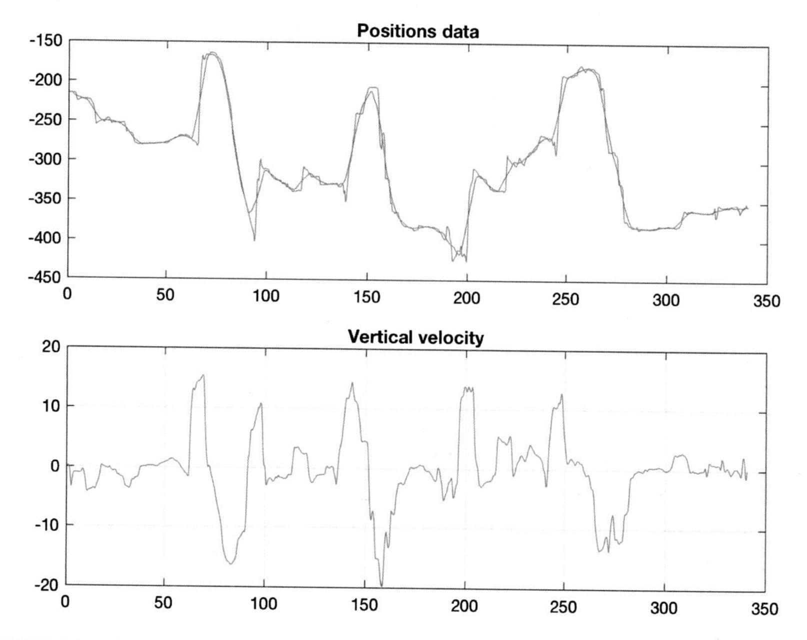

FIGURE 10.7 Surgical robot arm data: (top) vertical position (original in blue and smoothed in red), (bottom) vertical velocity.

  - These locations are nearly always arranged on a regular grid, and are often called *pixels*. The term pixel comes from 'picture element'. In the case of 3-D medical images, the term 'voxel' (volume element) is often used.
  - The image has a value (or intensity) associated with each pixel and we can visualize each pixel as a block of uniform color or a block of uniform shading (see Fig. 10.8).
- Discrete intensities:
  - As well as having discrete locations, the intensity value at each pixel is also usually discrete, i.e. it can take on one of a limited number of values.
- Data type:
  - The most common way to store the intensity of each pixel is using an integer data type, often between 0 and 255.
  - This is because a number between 0 and 255 can be represented by an unsigned integer type in a single byte (8 bits), which makes it easier to store intensities for many pixels in large images.
  - It is also possible to use other data types such as floating point numbers or similar.

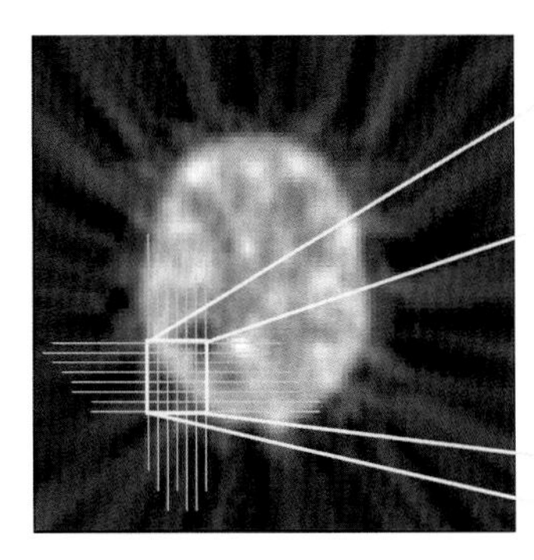
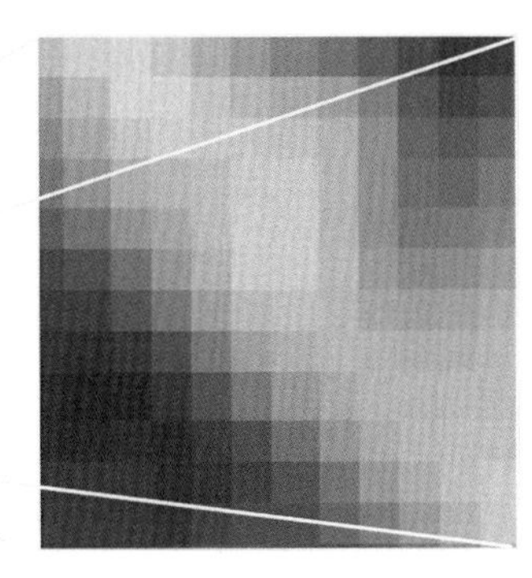

FIGURE 10.8 A PET image showing the individual pixels. These are the discrete locations at which we have a measurement.

10.6 ACCESSING IMAGES IN MATLAB

10.6.1 Color Versus Gray Scale Images

While everyday images tend to be color[1] images, medical images tend to be gray scale[2] images. A color image typically uses three numbers to represent the red, green and blue intensities associated with each pixel. Gray scale images only need to store a single value to represent a shade of gray between 0 (black) and a maximum value (white).

Storing pixel data for a gray scale image is relatively easy. The simplest way is to store an unsigned one-byte integer with 0 representing black and 255 (the maximum value) representing white. An integer between 0 and 255 will be shown as an intermediate shade of gray.

It is also possible for an image to have pixels that can be only black (minimum value) or white (maximum value), i.e. with no intermediate shades of gray. These are known as *binary* images.

Storing pixel data for a color image can be done in more than one way. Perhaps the simplest is to repeat the same approach used for gray scale images but with a separate number to represent the intensity of each of the red, green and blue (R, G and B) channels. In other words, we can use three bytes for each pixel, with one byte each for the red, green and blue intensities. For example, (0, 0, 255) would represent pure blue, (0, 255, 0) would represent pure green. Mixtures such as (255, 0, 255), which is magenta, are also possible. A shade of magenta with a lower intensity can be represented by (127, 0, 127) etc. Using a numeric value for each of the R, G and B intensities is sometimes called a *true color* representation.

---

[1] *colour* in the UK.

[2] *grey scale* in the UK.

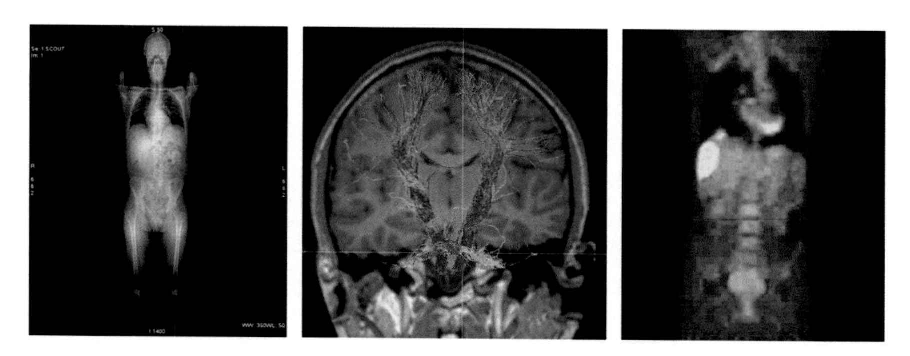

FIGURE 10.9 Medical images: (left) a CT scout image, (middle) a slice through an MR brain image (image generated using MRtrix: http://www.mrtrix.org/), (right) a PET scan.

Another way to store pixel data for a color image is to use a *color map*. A color map is a list of triples representing the RGB colors that appear in the image. In MATLAB, we can store a color map as an array with three columns. The number of rows will need to be sufficient to capture all the colors in the image. In order to represent an individual pixel, we only need to store a single integer which indexes the row in the map that contains the pixel's color. This method of using a color map and integers to look up rows in the map is called *indexed color* representation and it is useful for images with a reasonably limited range of colors.

10.6.2 Getting Information About an Image

We can use the built-in MATLAB function `imfinfo` to get information relating to an image file. First, we specify the name of the file, and then a `struct` variable (see Section 5.9) is returned which contains information on different aspects of the image, for example its size. There are three example images provided via the book's web site (see Fig. 10.9):

- `CT_ScoutView.jpg`: A CT scout image (i.e. an X-ray used to plan a CT scan) of a body.
- `mr-fibre-tracks.png`: A slice taken from a 3-D MR image of a brain, showing tracks of white matter in color.
- `pet-image-liver.png`: A PET image of the thorax and abdomen, showing metabolic activity in the liver.

■ Example 10.5

O10.A

Make sure that the images are in the current MATLAB path (see Section 3.7). We can use `imfinfo` to find information about an image as shown below.

```
>> imfinfo('mr-fibre-tracks.png')

          ans =
                 Filename: '/PATH/TO/FILE/mr-fibre-tracks.png'
              FileModDate: '08-Sep-2014 10:48:42'
                 FileSize: 192215
                   Format: 'png'
            FormatVersion: []
                    Width: 465
                   Height: 464
                 BitDepth: 24
                ColorType: 'truecolor'
           and some further information follows ...
```

We can see that the size of the pixel grid is 465 × 464, and that the image is a true color image. The `struct` has a `BitDepth` field which says that 24 bits (i.e. three bytes) are used for each pixel. ■

■ Activity 10.4

Use the `imfinfo` function to list the information for the other two provided images: `CT_ScoutView.jpg` and `pet-image-liver.png`. ■

O10.A

10.6.3 Viewing an Image

The simplest way to view an image from the command window is to use the `imshow` function,[3] giving the name of the image file as an argument.

■ Example 10.6

O10.A

```
>> imshow('mr-fibre-tracks.png')
```

The above call should result in the display shown below. Here, we see that `imshow` has loaded the image into a standard plotting window. The same tools that we can use for graph plots are also available in this window.

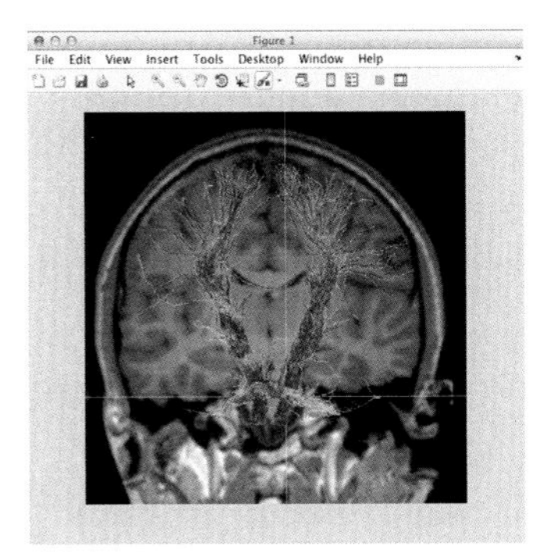

■

[3] The `imshow` function belongs to the Image Processing Toolbox. This toolbox is not part of the default MATLAB installation and needs a separate license. If a call to `imshow` fails, it may be because the toolbox is not licensed for your machine.

Another way to view an image is to use the *interactive* image viewer provided by the Image Processing Toolbox. This provides more ways to control how an image is viewed and also includes some tools for manipulating and editing the image. We can start it by calling `imtool` with the name of the image file:

```
>> imtool('mr-fibre-tracks.png')
```

This command loads the image into the interactive viewer with a *Tools* menu that offers some basic ways of interacting with the image. For example, under *Tools* → *Image Information* we can access the same information as we obtained using `imfinfo` above.

It is also possible to start the viewer without loading an image by simply typing `imtool` at the command window. This will open the viewer with a blank screen and an image can be loaded by going to *File* → *Open*...

■ Activity 10.5

O10.A

The images *CT_ScoutView.jpg*, *mr-fibre-tracks.png* and *pet-image-liver.png* are available on the book's web site. Try the different methods above for loading and viewing these images. ■

10.6.4 Accessing the Pixel Data for an Image

For simplicity, we will assume from now on that all images being described are gray-scale and that we can simply represent each pixel's intensity by one number. The example image *CT_ScoutView.jpg* is a single channel gray-scale image with an unsigned integer stored for each pixel. This means that we can store the intensities for all pixels in a single array. We obtain the pixel data using a call to the `imread` function[4]:

```
>> x = imread('CT_ScoutView.jpg');
>> whos
  Name          Size                Bytes  Class     Attributes
  x          1024x880              901120  uint8
```

Here, we load the pixel intensities into the array `x`. A call to `whos` tells us that the size of the array is 1024×880 pixels and its data type is `uint8`, i.e. unsigned 8 bit (1 byte) integers.

As we saw in Section 8.2.4, we can use the `imshow` function to display the pixel data in array `x` directly (i.e. without using the image filename). In other words, the command

```
>> imshow(x);
```

will produce the same result as `imshow('CT_ScoutView.jpg')`.

[4] `imread` is a built-in MATLAB function, not part of the Image Processing Toolbox.

10.6.5 Viewing and Saving a Sub-Region of an Image

We do not have to view the whole image, we can focus on a sub-region or 'window' inside the image. In the interactive viewer (`imtool`), we can use the zoom tool to adjust our view, zooming in and out as necessary.

We can also focus on a specific region using explicit commands. Picking out a sub-region is one of the simplest tasks we can carry out in image processing.

■ Example 10.7

As the image data are stored in an array, we can set ranges of row and column indices to specify the sub-region that we are interested in. For example:

O10.A, O10.C

```
>> x = imread('CT_ScoutView.jpg');
>> imshow(x(25:350,350:500))
```

This code displays a region of the CT scout image around the head and shoulders. Note that the rows specified in the example above (from 25 to 350) are counted starting from the top of the image. The columns (from 350 to 500) are counted from the left. ■

We can save our *cropped* version of the image data into a new image with the following command:

```
>> imwrite(x(25:350,350:500), 'ct-scout-crop.png')
```

This uses the built-in `imwrite` function that we saw in Section 8.2.4. Recall that this function can take two arguments: the array of image data to save (in this case a sub-region of `x`) and a string containing the file name to save it to.

■ Activity 10.6

Experiment with different crop regions and save or display the results. Use the command window and the tool built into the interactive `imtool` viewer. Can you choose a region which identifies the chest area in the CT scout image? ■

O10.A, O10.C

10.7 IMAGE PROCESSING

Now that we have seen the ways in which we can load gray scale image data, we will look at some simple operations that can be applied to modify the data.

10.7.1 Binarizing a Gray Scale Image and Saving the Result

Converting a gray scale image to a binary image produces an image with one of two values at each pixel (e.g. 'on'/'off', 'true'/'false', 0/1). We can use logical tests to do this in MATLAB. For example, using the same CT scout image described earlier:

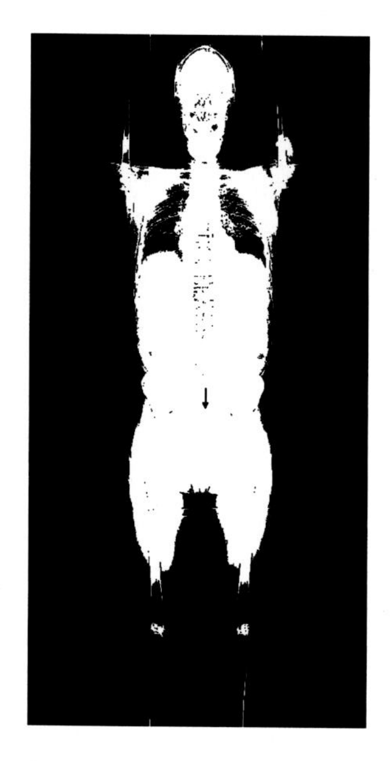

FIGURE 10.10 A binarized version of the CT scout image.

```
x = imread('CT_ScoutView.jpg');
y = x > 50 & x < 250;
```

This code first reads the image data into array `x`. The next line applies a pair of logical tests (element-wise) to `x` and applies the logical *AND* (`&`) operator to the results. This statement will produce an array (which is assigned to `y`) in which a value of `true` or 1 is present for those pixels in the original image with an intensity between 50 and 250 (exclusive).

A call to `whos` shows that `x` contains unsigned 8-bit integers and `y` is a logical array:

```
>> whos
  Name          Size              Bytes  Class      Attributes
  x          1024x880            901120  uint8
  y          1024x880            901120  logical
```

The resulting binarized image, displayed using `imshow(y)`, is shown in Fig. 10.10.

■ Activity 10.7

O10.A, O10.C

Generate your own binary image from the CT scout image and display it using your own ranges of values to decide which pixels are set to 0 and which are set to 1. ■

It is also possible to save the resulting array to a new image using the `imwrite` function.

```
% Save our binary (logical) data
>> imwrite(y, 'ct-binary.png')
```

We can now use `imfinfo` obtain general image information on the file that was saved. The salient parts of its output are given below:

```
>> imfinfo('ct-binary.png')

        ans =
               Filename: '/PATH/TO/FILE/ct-binary.png'
              % ...
                 Format: 'png'
                  Width: 880
              % ...
                 Height: 1024
               BitDepth: 1
              ColorType: 'grayscale'
              % ...
```

Note how MATLAB has saved the file to a gray-scale image that uses one bit for each pixel. This is because one bit is sufficient to indicate whether a pixel in a binary image is off (0) or on (1).

10.7.2 Threshold-Based Operations

Threshold-based operations take an image and update each individual pixel depending on whether its value is above (or below) a user chosen *threshold value*.

As a simple example, we can define a threshold value *tVal* and set any pixel's intensity to zero in the new image if the intensity at the same location in the old image is less than *tVal*, otherwise the value remains the same. At row i and column j, we denote the intensities of pixels in the old and new images by $old(i, j)$ and $new(i, j)$ respectively. The thresholding operation is then defined as:

$$new(i, j) = \begin{cases} 0 & \text{if } old(i, j) < tVal \\ old(i, j) & \text{otherwise} \end{cases}$$

We say the above operation 'thresholds to zero' (i.e. sets to zero any pixel below the threshold). In MATLAB we can easily do this with array operations.

■ Example 10.8

Here we set values to zero in the CT scout image if they are below 100.

O10.A, O10.C

```
x = imread('CT_ScoutView.jpg');
y = x;
tVal = 100;
```

```
y(x < tVal) = 0;
% View a sub-window of the
% input and output images.
imshow(x(25:350,350:500))
figure, imshow(y(25:350,350:500))
```

The last two lines display a sub-window of each of the input and output (old and new) images so that we can see the effect of applying the threshold. The resulting images are shown below.

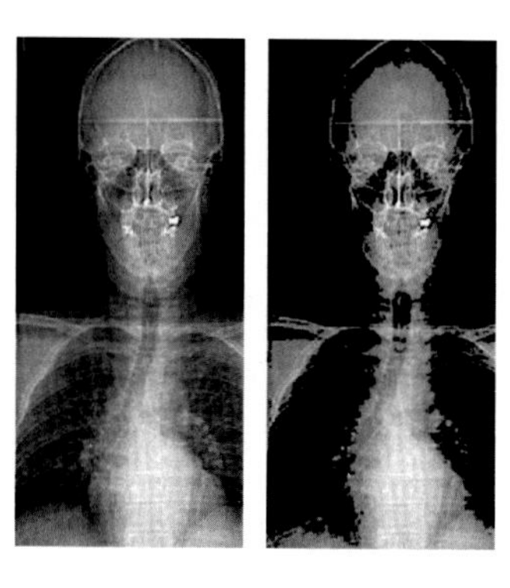

■

We can invert the way the threshold operation works, i.e. we can set to zero any pixel with a value *above* the threshold (rather than below). In other words:

$$new(i,j) = \begin{cases} 0 & \text{if } old(i,j) > tVal \\ old(i,j) & \text{otherwise} \end{cases}$$

We can alternatively use the threshold to limit or truncate the values in an image, for example by setting a pixel intensity to the threshold value if it was previously greater than the threshold:

$$new(i,j) = \begin{cases} tVal & \text{if } old(i,j) > tVal \\ old(i,j) & \text{otherwise} \end{cases}$$

The value assigned to a pixel above (or below) a threshold can be another choice for the user. In other words, as well as choosing a threshold value *tVal*, the user can set a substitute value *newVal* to apply to pixels above (or below) the threshold. For example:

$$new(i,j) = \begin{cases} newVal & \text{if } old(i,j) > tVal \\ old(i,j) & \text{otherwise} \end{cases}$$

■ Activity 10.8

O10.A, O10.C

Make your own threshold-based filter based on one of the types shown above. Apply it to the CT scout image and save or view the result. ■

10.7.3 Chaining Operations

We have only considered single operations so far but we can readily build up more complex ones by chaining them together. For example, we can apply more than one threshold operation to an image by making the output of one operation the input to the next. Again, we start with the gray scale CT scout image:

```
x = imread('CT_ScoutView.jpg');
y = x;
y(y < 50) = 0;
y(y > 200) = 0;
y(y > 0) = 255;
```

The first two lines read the image and copy it into an array variable `y`. The next three lines apply a chain of threshold operations to the array `y`, updating `y` on each line and modifying it again on the next.

■ Activity 10.9

Run the code above and compare it with the output of the binarizing code in Section 10.7.1. What do you notice? ■

O10.A, O10.C

This was a simple chain of operations but such chains can be made as complicated as we like. A chain of operations is often called a *pipeline*.

10.7.4 Image Data Type, Value Range, and Display

We need to be aware of some of the assumptions that MATLAB makes about the data types for images when reading and displaying them. For instance, in Example 10.7, we read data for a gray scale image into an array `x` and then used `imshow` to display (a part of) the image. If we inspect the contents of the variable `x` in this case, we can see that the data type was `uint8`.

When such an array, i.e. one with an integer data type, is passed to `imshow` an assumption is made that the full range of the image is from 0 to 255. This means that, when displayed, a value of 0 will correspond to a black pixel, a value of 255 will correspond to a white pixel and intermediate values will correspond to intermediate shades of gray. Any value above 255 is *clamped* and shown as a white pixel. Similarly, any value below zero is clamped and shown as a black pixel.

Sometimes, we may need to apply a processing operation to the image data that leads to an array with a floating point data type (e.g. `double`). If we then pass a floating point array to `imshow`, it will still make the assumption that black pixels correspond to a zero value. However, it will assume that white pixels correspond to a value of 1 (or any higher value). This can lead to unexpected results so care needs to be taken as shown in the example below.

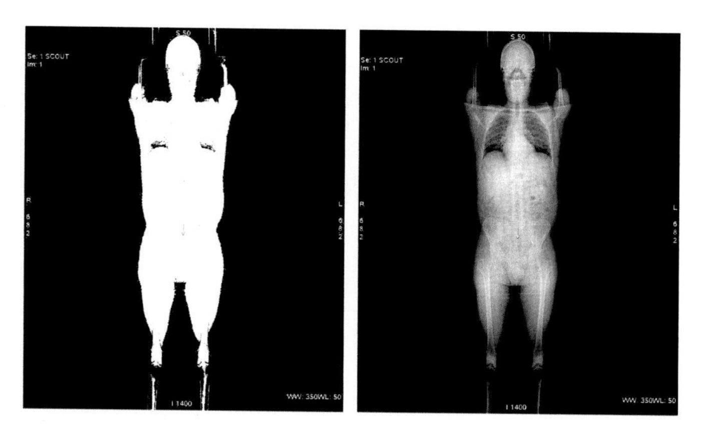

FIGURE 10.11 Displaying floating point version of a processed CT scout images: (left) without scaling to the range 0–1, (right) with scaling to 0–1.

■ Example 10.9

O10.A, O10.C

Hypothetically, assume that we need to take the square root of some data for an image. In the following, we load the CT scout image into an array `x` as before. Recall that `x` contains 8-bit unsigned integers (`uint8`). In order to take the square root of the array values, we first need to convert them to a `double` data type before passing them to the `sqrt` function. This is done in the second line. The result is stored in a variable `y` and then displayed in the last line.

```
x = imread('CT_ScoutView.jpg');
y = sqrt(double(x));
imshow(y)
```

In this case most of the non-zero pixels in the original image will now be shown as fully white pixels. This is because the original data ranged up to 255 and most non-zero pixel values, even after taking their square root, will still be greater than one. All such pixels will be shown as white by `imshow`.

In order to properly display such a `double` type array, we need to scale it so that it ranges from zero to one *before* passing it to `imshow`. This can be done in one line as follows.

```
imshow(y / max(y(:)))
```

■

The outputs from the two calls to `imshow` in the example above are shown in Fig. 10.11. On the left is the result of the first call (without scaling the floating

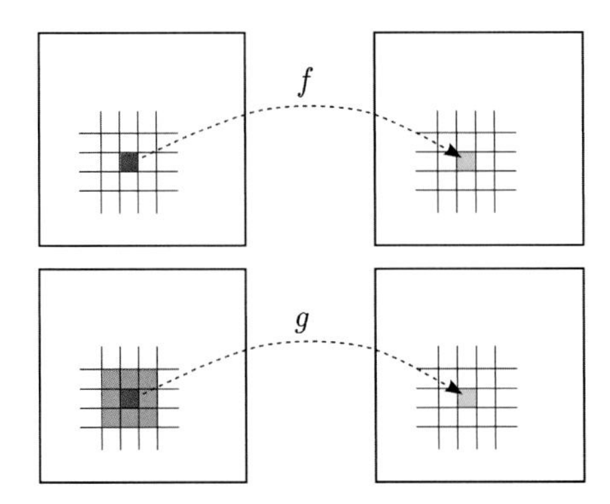

FIGURE 10.12 Top: a point processing operation in which the new pixel intensity depends only on the old pixel intensity. Bottom: a filter operation gives a new pixel intensity that depends on the old intensities in a neighborhood.

point square root data) and on the right is the result of the second call (after scaling to a range of 0 to 1).

10.8 IMAGE FILTERING

All of the operations applied to images that we have considered have been *point processing operations*. For such operations, the intensity of a pixel after the operation depends only on the same pixel's intensity before the operation. In general, for a point processing operation, we can write

$$new(i, j) = f(old(i, j))$$

for some function f. See Fig. 10.12 (top).

Filter operations are different because the new pixel intensity depends on all of the intensities in the original image that are in some *neighborhood* of the pixel. One simple neighborhood is a 3×3 window centered on the pixel. As illustrated in Fig. 10.12 (bottom), the pixel intensity in the output depends on all intensities in the neighborhood. Denoting the intensities in the 3×3 neighborhood as $\{v_1, v_2, \ldots, v_9\}$, we can write

$$new(i, j) = g(v_1, , v_2, \ldots, v_9)$$

for some function g that operates, in this case, on nine input intensities.

10.8.1 The Mean Filtering Operation

A very simple filtering operation is when we use the *mean* of the neighborhood values to obtain the new pixel value. That is, in the case of the 3×3 neighborhood, the function g above finds the mean of $\{v_1, v_2, \ldots, v_9\}$. In other words,

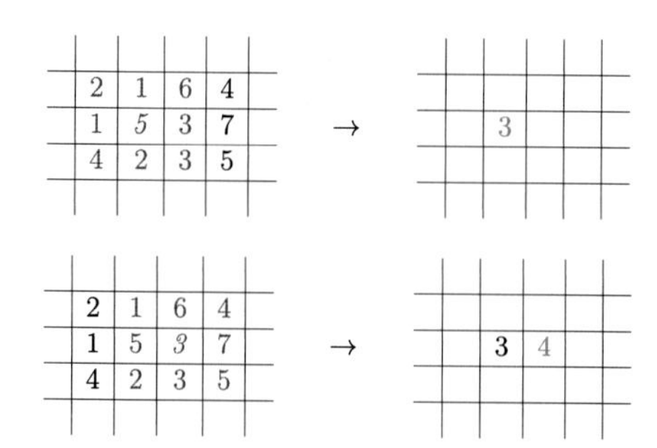

FIGURE 10.13 Computing new intensities using a mean filter.

for a 3 × 3 window,

$$new(i, j) = g(v_1, , v_2, \ldots, v_9) = \frac{1}{9}(v_1 + \ldots + v_9)$$

For example, the grid in Fig. 10.13 (top) shows a small part of an image. We assume that the image extends further around this part but we are ignoring it for now. The pixel with an intensity of 5 (in red italic) in the original image is assigned a new intensity of 3 because the mean of its neighborhood is 3.

$$\frac{1}{9}(2 + 1 + 6 + 1 + 5 + 3 + 4 + 2 + 3) = 3$$

To work out the new intensity for the next pixel, which has an original intensity of 3, we need to 'slide' the window by one pixel and recalculate the mean, as shown in Fig. 10.13 (bottom).

$$\frac{1}{9}(1 + 6 + 4 + 5 + 3 + 7 + 2 + 3 + 5) = 4$$

To calculate the new intensity in the next pixel, the window needs to slide again to cover a new neighborhood and the process repeated.

■ Activity 10.10

O10.C

The pixels at the border of the image do not have a 3 × 3 window of pixels that completely surround them. Decide on a strategy for dealing with these border pixels when applying a mean filter. More than one strategy is possible. ■

10.8.2 The Actual Filter Used

If we expand the calculation for the pixel shown in Fig. 10.13 (bottom) it can be written as:

$$\frac{1}{9} \times 1 + \frac{1}{9} \times 6 + \frac{1}{9} \times 4 + \frac{1}{9} \times 5 + \frac{1}{9} \times 3 + \frac{1}{9} \times 7 + \frac{1}{9} \times 2 + \frac{1}{9} \times 3 + \frac{1}{9} \times 5$$

...	$\frac{1}{9}\times 1$	$\frac{1}{9}\times 6$	$\frac{1}{9}\times 4$	
...	$\frac{1}{9}\times 5$	$\frac{1}{9}\times 3$	$\frac{1}{9}\times 7$	
...	$\frac{1}{9}\times 2$	$\frac{1}{9}\times 3$	$\frac{1}{9}\times 5$	

FIGURE 10.14 Applying a mean filter in 2-D.

i.e. the mean filter produces a new pixel intensity by taking a weighted sum of the intensities in the original image.

The weights can be arranged in a 3×3 array which we call the *filter*,

$$\begin{pmatrix} \frac{1}{9} & \frac{1}{9} & \frac{1}{9} \\ \frac{1}{9} & \frac{1}{9} & \frac{1}{9} \\ \frac{1}{9} & \frac{1}{9} & \frac{1}{9} \end{pmatrix}$$

In exactly the same way that we did for the convolutions in 1-D in Section 10.4, we can view this filter as sliding over the image and being used to weight and sum the corresponding pixel intensities (see Fig. 10.14). The only difference is that the filter slides in *two* directions (horizontally and vertically). For instance, we can slide the filter horizontally across a row from left to right until we reach the end of the row. We can then restart the process on the next row down and so on.

In this case, the operation is a little trivial because all of the weights are equal (to 1/9). But we can vary the weights to obtain different effects. For example, we can use a filter that gives a higher weight to the central pixel in the neighborhood and lower weights to the others:

$$\begin{pmatrix} \frac{1}{12} & \frac{1}{12} & \frac{1}{12} \\ \frac{1}{12} & \frac{1}{3} & \frac{1}{12} \\ \frac{1}{12} & \frac{1}{12} & \frac{1}{12} \end{pmatrix}$$

10.8.3 Applying a Filter in MATLAB

We can apply a filter to an image in MATLAB with the function `imfilter`, which is part of the Image Processing Toolbox. In the case of the mean filtering operation above, we saw that we had the very simple filter of a 3×3 array with all entries equal to 1/9.

■ Example 10.10

This code shows a filter of uniform values of $\frac{1}{9}$ being created by starting with an array of ones and dividing by 9. This is then applied to a sub-region of the CT image using `imfilter`. The `imfilter` function does the job of 'sliding'

O10.A, O10.C

the filter over each pixel in the image and performing the calculation of the weighted sum. The input and output images are shown below.

```
imData = imread('CT_ScoutView.jpg');
% sub-region
reg = imData(50:120, 400:450);
% Create the mean filter.
myFilt = ones(3) / 9;
% Apply the filter
filtReg = imfilter(reg, myFilt);

figure
imshow(reg);
figure
imshow(filtReg);
```

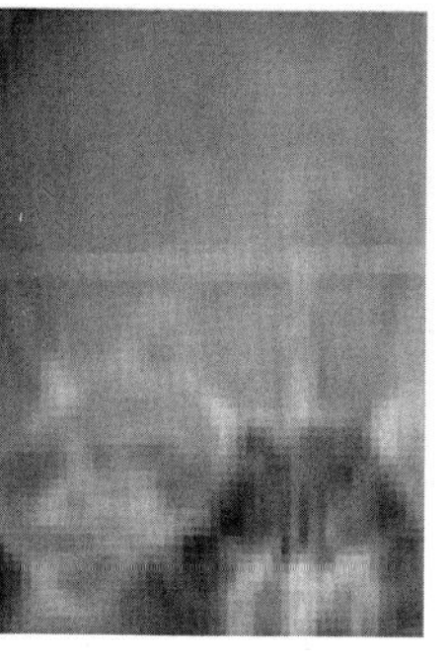

■

The visualization in Example 10.10 shows the image region before and after applying the filter. Note how the result after the mean filter appears smoother and less sharp than the original. If you look carefully, you should also notice a darker region around the edges of the new image. At these positions, the filter only partly overlapped with the original image.

The step in the commands above which assigns a filter to the variable `myFilt` is easily replaced with a different array assignment if we want to use a different filter. The subsequent call to the `imfilter` function would still be the same.

■ Activity 10.11

O10.A, O10.C

Make your own 3 × 3 array containing a filter. Apply it to the data array of a gray scale image using the `imfilter` function. Perhaps you can make a filter with a higher weight for the central pixel and lower weights for the remainder. Experiment with filters whose elements sum to one or a value different from one. Visualize or save the results. ■

■ Activity 10.12

O10.C

Write your own implementation of the `imfilter` function which again takes two arguments: an array of gray scale image data and a square array containing the filter to apply. You can assume that the filter array is always a 3×3 array to make things simpler. It is possible to carry this out as a series of nested loops with the following structure:

```
% ...
% pair of loops for whole image array.
for row = 2:nrows - 1
  for col = 2:cols - 1
    % Current pixel has indices 'row' and 'col'

    % Pair of loops for local window
    for rowOffset = -1:1
      for colOffset = -1:1
        % Do stuff here
        % to get a new value
      end
    end % end of window loops

    % May need to assign here.
  end
end % end image loops
```

■

10.8.4 Filtering and Convolution

You should have noticed that the filtering operations described in then previous sections have a lot in common with the convolution operations described for 1-D signals (Section 10.4). What we have described as a filter can also be called a *kernel*. The process of 'sliding' the filter/kernel over the image to produce the output is directly analogous to the sliding we looked at in the case of 1-D signals. The only difference in the case of images is that this sliding operation needs to take place in two dimensions, along both rows and columns. In the 1-D case, the kernel is a row vector (1-D array) whereas in the 2-D case the kernel is a matrix (2-D array).

MATLAB provides a 2-D convolution function called `conv2` and it can be used instead of the `imfilter` function. The caveat is that it requires and returns data of type `double` so care needs to be taken when passing the data and visualizing the result.

■ Example 10.11

O10.A, O10.C

Recall the `imfilter` function we used to apply a mean filter in Example 10.10. Here, we will use `conv2` instead. Note how the data are converted to `double` before passing to `conv2`. Note also how the data are scaled to the 0–1 range by dividing by 255 before visualization.

```
imData = imread('CT_ScoutView.jpg');
```

```
myFilt = ones(3) / 9;

reg = imData(50:120, 400:450);

filtReg2 = conv2(double(reg), myFilt, 'same');
imshow(filtReg2 / 255.0)
```

Note also how the call to `conv2` was given the argument `'same'` to ensure that the output array was the same size as the input array. This is again analogous to the similar call in the 1-D case (see Example 10.2). ■

As was the case with the 1-D convolution, one detail relates to the flipping of the kernel when carrying out the convolution. In the case of `conv2`, the kernel is flipped left–right and up–down before being passed over the image array. In the case of Example 10.11, these flips had no effect as the kernel was symmetric. In general, if the Image Processing Toolbox is available, the use of `imfilter` is a lot easier than the use of `conv2`.

10.9 SUMMARY

In this chapter we have looked at signals and images and how we can read, process and visualize them using MATLAB commands. We have examined the convolution operation and how it can be used to process 1-D and 2-D signals. We have also looked at some of the properties of images such as discrete locations and gray scale versus color images. Focusing on gray scale images, we have introduced some of the basic steps that are used for *image processing* such as threshold-based operations. These operations rely mainly on the fact that, once an image is loaded into MATLAB, the data it contains can be represented by an array variable that can be manipulated to process the data. We have used both standard built-in MATLAB functions as well as functions provided by the more specialist Image Processing Toolbox. We have discussed image filtering and seen its close relationship to the concept of convolution of 1-D signals. While these basic steps can be applied to images in general, they form the basis of the important field of *medical image processing* which has been used in a wide range of clinical settings since a variety of different types of medical image are now routinely available. The basic operations we have looked at can be 'chained together' with the output from one step forming the input for a later step. In this way image processing 'pipelines' can be built up incrementally.

10.10 FURTHER RESOURCES

- The Image Processing Toolbox has a dedicated help section that is distinct from the main MATLAB help. It has sections including:
  - *Import/export*: Reading images into MATLAB and saving them.
  - *Display*: Showing an image through the interactive viewer or by a command line call to a function such as `imshow`.

- For readers interested in more details, there is a Signal Processing toolbox available in MATLAB with its own help section as above. It also provides some interactive tools.
- An interactive tutorial on Signal Processing is available at the MathWorks web site: https://www.mathworks.com/academia/student_center/tutorials/signal-processing-tutorial-launchpad.html.
- A number of examples using image processing are also available in the online MATLAB documentation: http://www.mathworks.com/help/images/examples.html?s_cid=doc_ftr.

EXERCISES

These exercises involve use of a number of sample 1-D signals and medical images which you will need to download from the book's web site before proceeding.

■ Exercise 10.1

O10.A, O10.B

In Fig. 10.1 (top) we saw some example ECG signal data. The data for this plot are available on the book's web site, in the file called *example_ecg_data.txt*. Write code to load the data from the file and replicate the plot shown in Fig. 10.1.

The data in the file were sampled at 720 Hz. Write code that down-samples the data to rate of 60 Hz and plots the down-sampled version of the data overlaid onto the original data. Ensure your plot is annotated and contains a legend. ■

■ Exercise 10.2

O10.A, O10.B

A common kernel for filtering a signal for smoothing is the *Gaussian* kernel. This is based on the Gaussian distribution (also known as the normal distribution). The continuous Gaussian function is shown below (left) and values taken from this function at discrete points may be taken to form discrete Gaussian kernels. The center plot shows three values taken from the function and the right hand plot shows five values being taken. In each case, the values may be normalized so that they sum to one in order to generate a kernel.

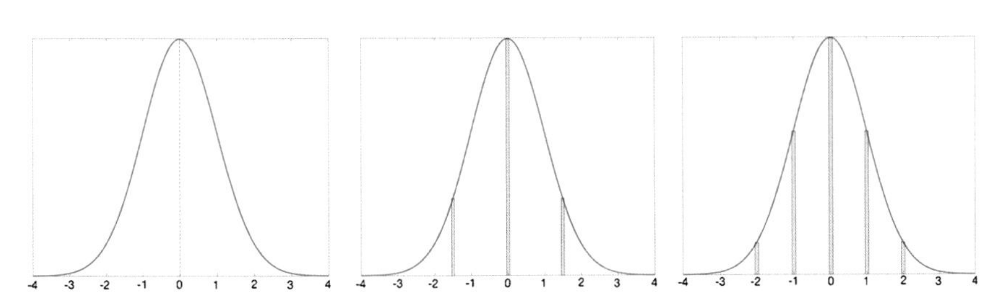

The Gaussian kernels obtained as described above, of lengths three and five are

[0.2, 0.6, 0.2]   [0.06, 0.24, 0.40, 0.24, 0.06]

Continuing the work of Exercise 10.1, use the `conv` function to smooth the down-sampled (or *subsampled*) ECG data using both the Gaussian kernels provided above. Plot the subsampled data and both versions of the smoothed data in the same window. ■

■ Exercise 10.3

O10.A, O10.B

The data file for one of the sets of heart rate data shown in Fig. 10.1 is called *'heart_rate_series_1.txt'* and is available to download.

Write code to read in these measurements and use the `smooth` function to reduce the level of noise in the data. Write code to find the first time derivative of this heart rate signal (i.e. find the rate of change of the heart rate). See Example 10.4 to get a hint on how to do this.

Make a plot showing the first 15 seconds of the smoothed data in the top half of the window and the time derivative in the bottom half (*hint:* see the documentation for the `subplot` command). Experiment with different values of the span for the `smooth` function to see the effect on the estimated derivative. Remember that the derivative estimates at the end may not contain sensible values so it is safe to set these to zero. ■

■ Exercise 10.4

O10.A, O10.C

Using the *CT_ScoutView.jpg* image, produce, display and save a binarized image of the head using an appropriate threshold (i.e. see the figure below).

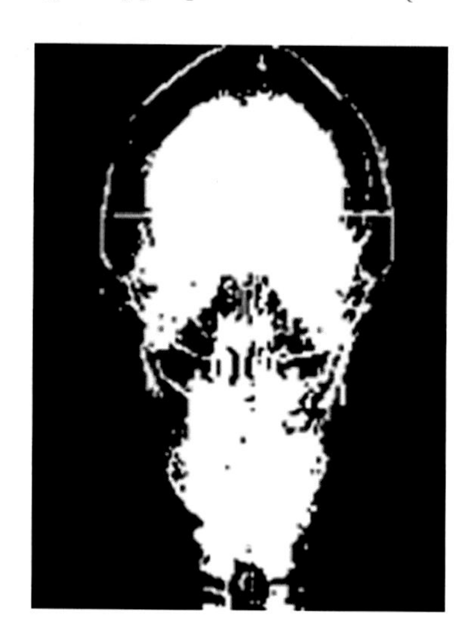

■

■ Exercise 10.5

Write two functions called `thresholdLow` and `thresholdHigh`, which both take an image and a threshold as their two arguments, and produce a thresholded image as their result (the thresholded images should have zero intensity at any pixel that had a value below/above the threshold in the original image). ■

O10.A, O10.C

■ Exercise 10.6

Write two functions called `truncateLow` and `truncateHigh`, which both take an image and a threshold as their two arguments, and produce a truncated image as their result (the truncated images should have the intensity of any pixel below/above the threshold set to the threshold). ■

O10.C

■ Exercise 10.7

Write a pipeline of operations to process *pet-image-liver.png* to highlight the area of high activity in the liver, i.e. the output of the program should be as shown in the figure below.

O10.A, O10.C

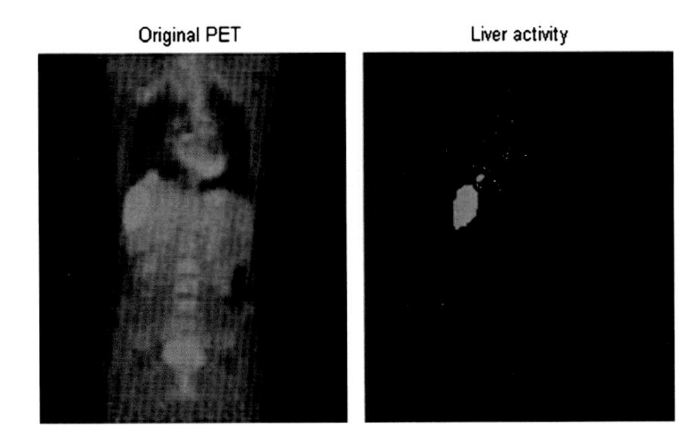

(Also display a count of the number of pixels that are non-zero in the final image.) ■

■ Exercise 10.8

In binary image processing, it is possible to apply a *dilation* operation to an image. You are provided with a simple binary image showing a map of cortical gray matter in the brain. This is saved in the file *cortex-gm.png* and is shown below.

O10.A, O10.C

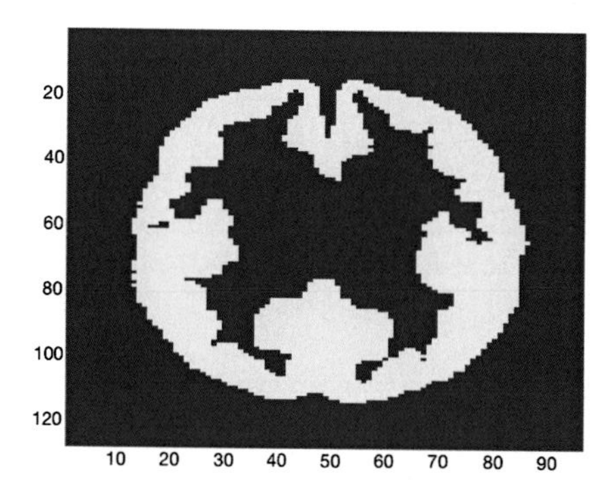

The process of dilating a binary image converts some of the background (blue) pixels to foreground (yellow) pixels. A background pixel is 'switched on' if one of its neighbors (up, down, left or right) is a foreground pixel. Dilating the original image gives the image shown below.

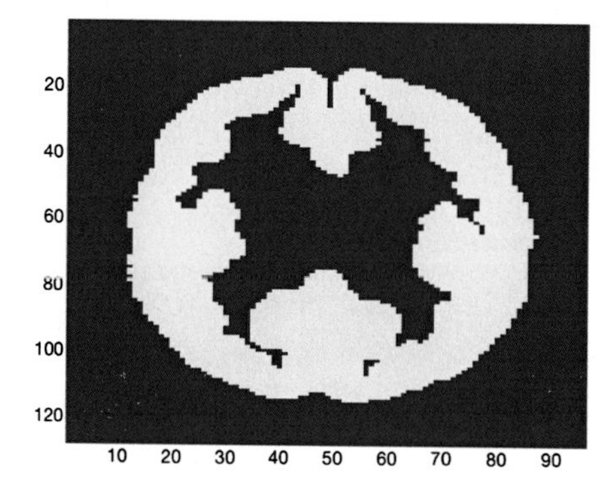

Write a function that accepts an image matrix as an input argument and returns a new matrix as an output containing the dilated version of the input. Write a script *m*-file that loads in the image, calls the function to dilate the image, and then displays both the original and dilated images. ■

FAMOUS COMPUTER PROGRAMMER: BJARNE STROUSTRUP

Bjarne Stroustrup is a Danish computer scientist, born in 1950 in Aarhus, Denmark. He studied at Aarhus University in Denmark and Churchill College, Cambridge, in the UK where he got his PhD in Computer Science.

Stroustrup is famous for developing the C++ programming language. C++ is responsible for popularizing the concept of Object-Oriented Programming (OOP). Before C++, most programming languages were 'procedural', i.e. programs consisted of a sequence of steps to run an algorithm. OOP models problems as a set of objects that contain data along with procedures for operating on that data; In a running OOP program, objects send each other messages and data (see Section 1.1.1). OOP is now an important paradigm in the world of computer programming. Stroustrup developed C++ in 1978 as an extension to the popular C programming language (see Chapter 7's Famous Computer Programmer). In his own words, he "invented C++, wrote its early definitions, and produced its first implementation ... chose and formulated the design criteria for C++, designed all its major facilities, and was responsible for the processing of extension proposals in the C++ standards committee."

Stroustrup currently lives in Texas, where he is a Professor of Computer Science at Texas A&M University.

> *"An organization that treats its programmers as morons will soon have programmers that are willing and able to act like morons only."*
>
> **Bjarne Stroustrup**

CHAPTER 11

Graphical User Interfaces

LEARNING OBJECTIVES

At the end of this chapter you should be able to:

O11.A Create and design a graphical user interface (GUI) and add components to it

O11.B Appreciate that user interactions with a GUI trigger events that lead to specific callback functions to be invoked

O11.C Use callbacks and other functions that allow a user to interact with a GUI and its components

11.1 INTRODUCTION

In this chapter, we discuss how to create graphical user interfaces (GUIs) using MATLAB. In the preceding chapters, we have been writing MATLAB programs that interact with the user by displaying text to the command window and reading data from the keyboard, but GUIs have an important role to play in making our programs more accessible and usable by a wider range of (non-technical) users. An outline of how to build a simple GUI with basic components such as text boxes or axes is given using the built-in `guide` tool. We also look at how to control the behavior of GUI components using handles and callback functions. For more information on handles and how they are used, refer to Section 11.3 the end of the chapter.

11.2 GRAPHICAL USER INTERFACES IN MATLAB

Whilst a lot of interaction with a software environment can be carried out on a command line or by running a script, it can also be useful to interact via a *graphical user interface* or GUI.

In MATLAB, we can create our own custom-made GUIs, we can add objects including text-boxes for the user to type into, pop-up menus for making choices, images or plots, and buttons to turn something on/off. We can customize the

MATLAB Programming for Biomedical Engineers and Scientists. DOI: 10.1016/B978-0-12-812203-7.00011-2

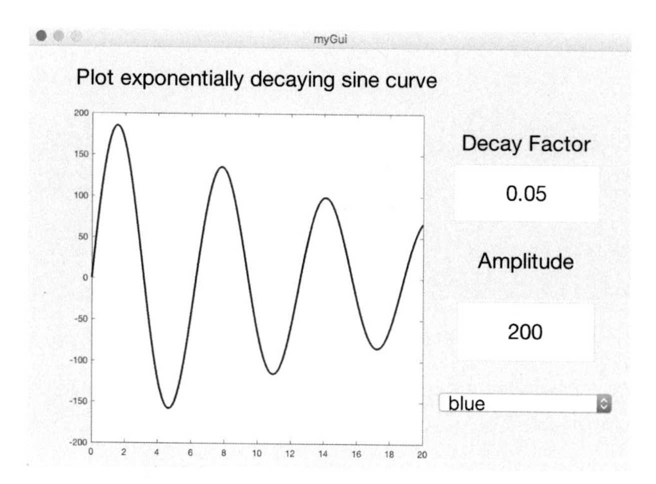

FIGURE 11.1 An example of a GUI. This displays a plot and has some components (text boxes and a drop down menu) to allow the user to interact with the plot.

GUI, laying out the components and sizing them as we want. An example of such a GUI is shown in Fig. 11.1.

When a user interacts with a component, such as clicking a button or entering some text, this is called an *event*. When an event happens, MATLAB executes a specific *callback function* that is associated with the event. We can edit the callback function to control the behavior of the GUI depending on what the user did.

A single component can have more than one callback function, typically one for each different type of event (i.e. action by the user). For example, a set of axes (say the one in the plot above) can have a callback function for when the user presses down the left mouse button and a different callback function for when the mouse pointer simply moves inside the axes.

11.2.1 Building a GUI with the Guide Tool

It is possible to create and build a MATLAB GUI entirely in code using scripts and functions. However, in practice, it is a lot easier to use the *Graphical User Interface Development Environment*, which is shortened to '`guide`'.

We will illustrate the use of the `guide` tool for building a GUI with a series of examples that build up a simple GUI step by step. This GUI will carry out some basic plotting of data in a way that can be modified by the user.

■ Activity 11.1

O11.A **Creating a blank GUI and saving it:** To start the tool, simply type `guide` at the command window. This causes a dialog box to be displayed (see

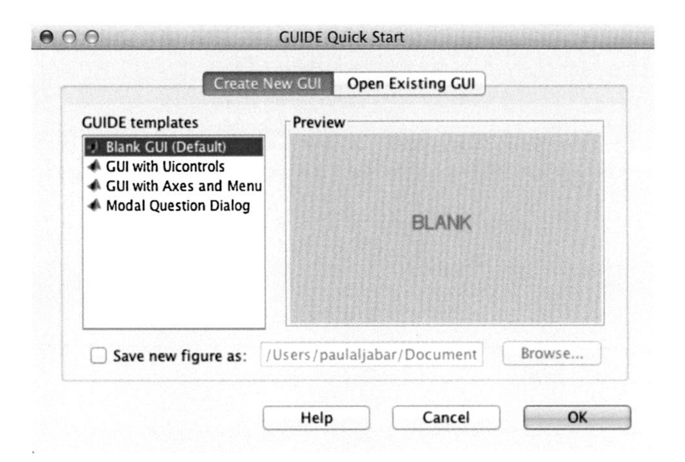

FIGURE 11.2 Starting the `guide` tool.

Fig. 11.2) in which we can choose the type of GUI to build. Choose the default option of starting a blank GUI. This opens a blank canvas for our GUI which we can fill in as we like (see Fig. 11.3). ■

The blank canvas is overlaid with a grid to help us with the layout of the components (buttons, text-boxes, etc.) that are inserted. There is a set of buttons on the left that we can use when editing. The button with the arrow icon at the top is used for selecting components and moving them around. Below that, there are buttons for various components that can be placed in the GUI. Hovering the mouse over a button gives some text to describe what the component is. For example, the button that has an icon showing a small graph of a function is used to insert a set of axes into the GUI for plotting graphs or displaying images.

■ Activity 11.2

Our GUI is still untitled so we will save it first before carrying on. Click *File* → *Save As* to open the save dialog box. Because there will eventually be more than one file used by the GUI, it is a good idea to save the figure in a folder of its own where it can be kept together with any other files needed. For example, save it as *myGui.fig* to a new folder called *myGui* (see Fig. 11.4). ■

O11.A

After the save step, there will be two files in the folder, *myGui.fig* and *myGui.m*. The first is a binary file containing our GUI's window and components, and the second is a standard MATLAB *m*-file that will contain our callback functions.

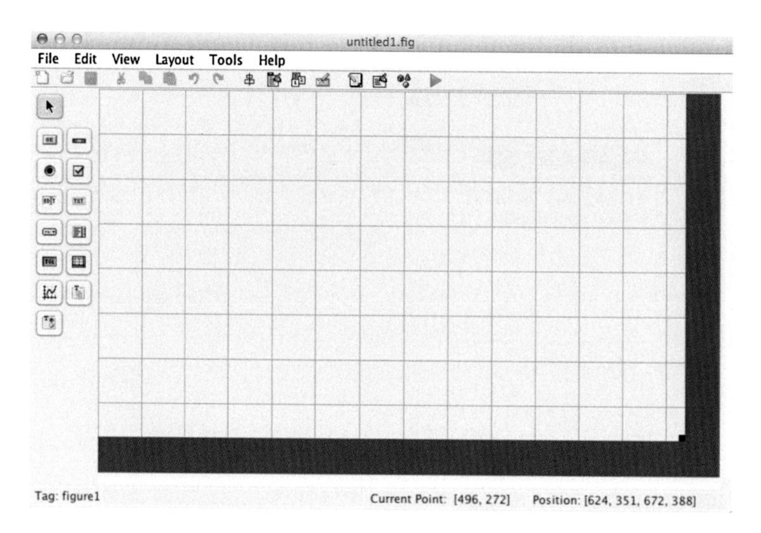

FIGURE 11.3 A blank canvas in `guide`.

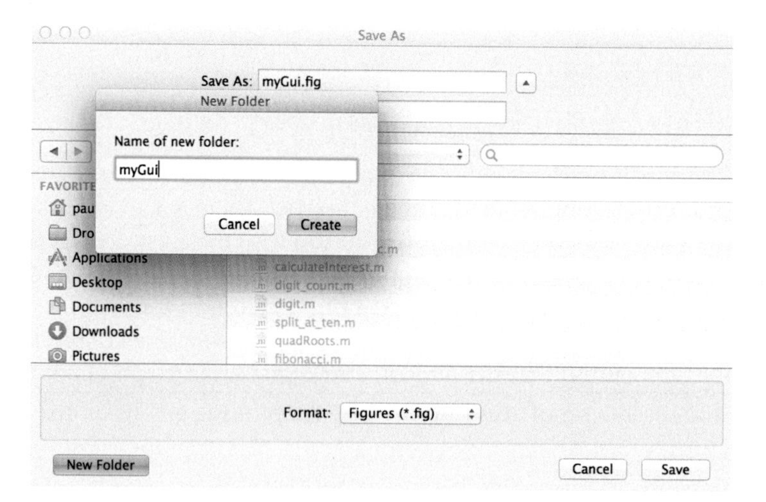

FIGURE 11.4 Saving a GUI using `guide`.

> **Automatic code:** Note that, after saving, MATLAB opens an editor window for the *myGui.m* file with a lot of complex looking code. This code is automatically created by the `guide`.

It is now possible to run the GUI we have just created, even though it does not contain any components yet. This can be done by clicking the *Run* button (▷) at the top of the script editor window as we would for any MATLAB *m*-file. MATLAB may complain about the script not being in the path, in which case,

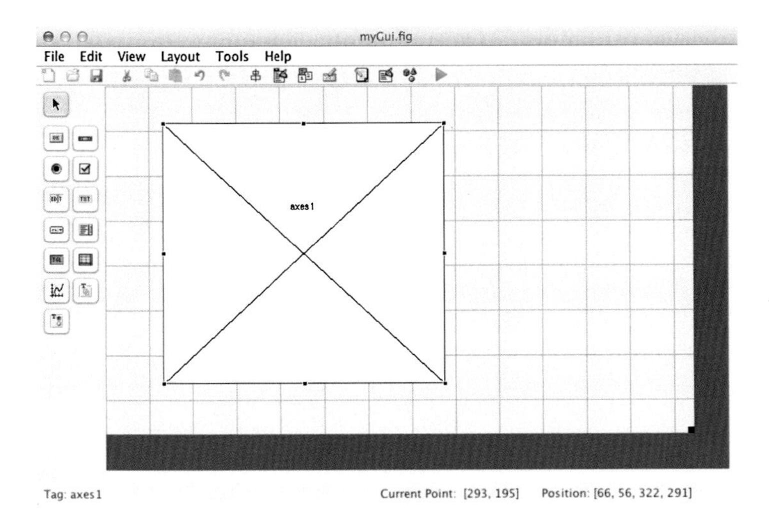

FIGURE 11.5 A `guide` GUI with axes.

choose the option to *Change Folder* so that the GUI can be run. This changes the MATLAB current directory to the location of *myGui*. When the GUI runs, as expected, we get a blank window that we cannot interact with yet. We can close the window to continue editing.

The GUI can also be run directly from the `guide` canvas screen which also has a green *Run* button at the top. Finally, it can also be run by typing `myGui` at the command window (as long as its folder is in the path or we have changed the working directory to that folder).

■ Activity 11.3

Adding a component to the GUI: Now we can add components to our GUI. We would like to include a plot, so we start by adding a set of axes. Click on the button to insert axes. This makes the cursor turn into a cross-hair. Drag out a square or rectangle on the canvas to generate a set of axes. The GUI canvas should now look something like Fig. 11.5.

O11.A

Move and drag the corners of the axes until they have the size and position required. Leave some space to the right of the axes for some components to be added later. ■

Re-opening at a later stage: If we close our MATLAB session and later, in a new session, we need to re-open our GUI in design mode, we can just type `guide` at the command window. We should then choose to *Open an existing GUI* and, if our GUI is not already listed, select *Browse* in order to locate to the directory where it was saved.

11.2.2 Controlling Components: Events and Callback Functions

Now that we have a component added to the GUI (a set of axes), we can modify the code that controls components and their behavior. In MATLAB, and in some other programming languages that allow GUI programming, this can be done using *callback functions*.

A callback is special function that is associated with each component on the GUI. Callbacks are linked to *events* that occur when the user interacts with the GUI, e.g. by clicking on an item in the GUI (in MATLAB this event is called *ButtonDown*). Other less obvious events include the 'creation' of each of the components as the GUI is starting up or their 'deletion' when the GUI closes.

When an event occurs, the system will execute code in the specific callback function associated with that event. We can modify this function, inserting our own code, in order to control the GUI's behavior.

■ Activity 11.4

O11.B

Accessing the callback function for a component: Now that we have a component (some axes) on our GUI in the `guide` editor, we can have a look at what callback functions it has available. We can do this by right-clicking on our component (the axes) to bring up a context menu. As shown in Fig. 11.6, we can *View callbacks* in this menu. Callback functions can also be accessed from the *View* menu at the top of the `guide`.

In this case, we have three functions available as callbacks: `CreateFcn`, `DeleteFcn` and `ButtonDownFcn`. These are associated with the creation, deletion and button-down events for the axes.

If we click on the first of these in the menu (`CreateFcn`) then the `guide` will automatically generate code for this function and will switch focus to this code in the GUI's *m*-file. ■

The automatically generated code should look like the following:

```
% Executes during object creation, after setting all properties.
function axes1_CreateFcn(hObject, eventdata, handles)
% hObject
%    handle to axes1 (see GCBO)
% eventdata
%    reserved - to be defined in a future version of MATLAB
% handles
%    empty - handles not created until after all CreateFcns called

% Hint: place code in OpeningFcn to populate axes1
```

This specially generated code is run when the axes are created as the GUI is being loaded. The name of the function reflects the fact that the axes we inserted have been given the label *axes1* and hence the function is named

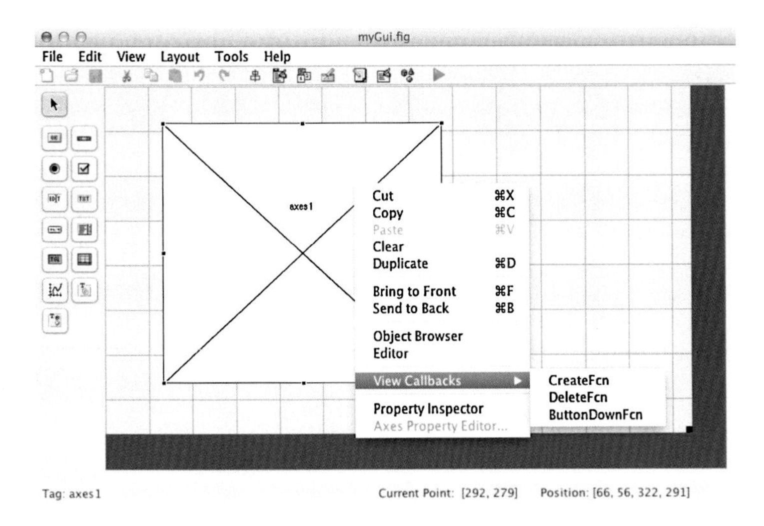

FIGURE 11.6 Accessing the callback functions in the `guide` editor.

`axes1_CreateFcn`. This would distinguish the callback from a callback that would be needed for another set of axes, if they were later added to the GUI.

The usage text for the function tells us that the first argument is an object *handle*[1] to our axes (`hObject`), and that the second argument can be ignored. It then says that the third argument (`handles`) is currently empty and not usable. We will look further at this particular `handles` variable later on when we consider how to pass information between the various functions.

From this relatively uninformative start, the last line of the usage gives us a hint to add code to a specific function in order to populate our axes. This is what we shall do next.

Adding Code to Populate the Axes

The hint in the usage text for the axes creation code above suggests that we should add code to the GUI's *opening* function and this is what we will do now. The GUI's opening function is in the same *m*-file and its initial code has also been automatically generated. We can figure out the name of the function we are looking for. It is named `myGui_OpeningFcn` as it is associated with the whole GUI (*myGui*) and it is the opening function (i.e. just before it is made visible). Currently, this function mostly contains comments describing what it does:

---

[1] A handle is a way of referring to a more complex object in MATLAB. See Section 11.3 for a more general description of handles.

```
% --- Executes just before myGui is made visible.
function myGui_OpeningFcn(hObject, eventdata, handles, varargin)
% This function has no output args, see OutputFcn.
% hObject    handle to figure
% eventdata  reserved - to be defined in a future version of MATLAB
% handles    structure with handles and user data (see GUIDATA)
% varargin   command line arguments to myGui (see VARARGIN)

% Choose default command line output for myGui
handles.output = hObject;

% Update handles structure
guidata(hObject, handles);
```

■ Activity 11.5

O11.B

We can inspect the opening function's behavior by debugging the code. Place a breakpoint on the last line of code (i.e. `guidata(hObject, ... handles);`) and press the *Run* button. The debugger should stop at this point in the code before the GUI is made visible.

There is only really one variable of interest at this point, the `handles` variable. Type its name at the command window and hit return. The output should be similar to the following:

```
>> handles
handles =
    figure1: 173.0077
      axes1: 174.0077
     output: 173.0077
```

■

Note that `handles` is a `struct` (i.e. a structure data type, see Section 5.9). `handles` has three fields, *figure1*, *axes1* and *output*. Each of these fields is a handle to the object that its name describes. Our axes can be accessed through the second of the handles, and we can plot to them by referring to their handle. To do this, we can use a form of the `plot` command where we specify the axes in which we want to plot with the first argument.

Now that we have seen how the components are represented in a structure containing handles, we will next use the handle of the axes to produce the plot.

■ Activity 11.6

O11.B, O11.C

We will plot some of the ECG data that we first saw in Fig. 10.1 and later in Exercise 10.1. The data for this plot are available on the book's web site, in the file called *example_ecg_data_b.txt*. Save a copy of this file into the same folder that contains the code (*.m*) and figure (*.fig*) files for the `myGui` GUI.

Stop the debugger and add code to the `myGui_OpeningFcn` function so that it looks like the following.

```
% Choose default command line output for myGui
handles.output = hObject;

% START ADDED CODE
% Read data
ecg = csvread('example_ecg_data_b.txt');
nPts = length(ecg);

% Sampling frequency
sampFreq = 720; tStep = 1/sampFreq;
% Time points at given sampling frequency.
t = 0:tStep:(nPts-1)*tStep;

plot(handles.axes1, t, ecg);  % Plot data:
% END ADDED CODE

% Update handles structure
guidata(hObject, handles);
```

The added code section is marked by comments at the start and the end. Note how the `plot` command takes `handles.axes1` as the first argument. ■

Now when we run the GUI, our axes will contain a plot (see Fig. 11.7). Note how, for this ECG data file, we have measurements extending over five seconds. One way to introduce more user interaction is have a parameter that we can vary and control and which determines the portion of the data that we plot. For example, the user may want to set a maximum value for the time and the plot's horizontal axis can be adjusted to reflect this. We will use a text box to get this user input.

Using a Text Box to Vary the Plot

There are two components in the `guide` for inserting text boxes into a GUI: *Edit Text* and *Static Text*. The first inserts a text box that the user can type into (including numeric values) and the second cannot be modified by the user and simply displays some text in the GUI.

We can put both kinds of text into our GUI. We can place an edit text box to get the upper limit of the time range for plotting the ECG data. We can also add a static text box to act as a label for the edit text box.

■ Activity 11.7

Using the `guide` editor, place a static text box and an edit text box onto the canvas. By default the static text component will just display the string '*Static Text*' and we need to change this. Click to highlight the component and, from the *View* menu or the (right-click) context menu, choose *Property Inspector*. Find the *String* field and change it from '*Static Text*' to '*Upper Time Limit*'.

O11.A

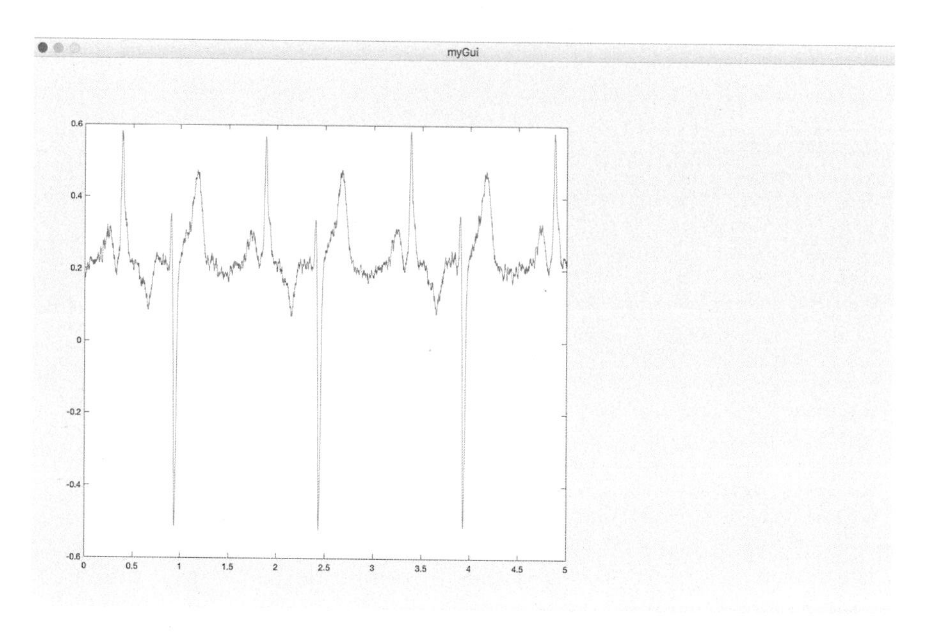

FIGURE 11.7 The `guide`-created GUI containing a plot of some ECG data.

Now highlight the edit text component and open its property inspector. Find the *Background Color* field and ensure that it is set to white. This helps to emphasize that it can be edited by the user. Then go to the *String* field and delete the *'Edit Text'* that is there by default so that the box is empty. Select *File* → *Save* on the `guide` or click the *Save* icon. The GUI canvas should now look something like Fig. 11.8. ■

Finding Out Which Callbacks Are Available for Text Boxes

Not all components on our GUI will have all the same callback functions. We can use the `guide` editor to identify which callbacks are available for individual components. The easiest way to do this is to right-click on the component and open up the menu item that lists the callback functions.

■ Activity 11.8

O11.B

In the `guide` editor, right-click on each of the text boxes to see which callback functions are available. ■

You should have found that both types of text box (edit and static) have the following callback functions:

- `CreateFcn`
- `DeleteFcn`
- `ButtonDownFcn`

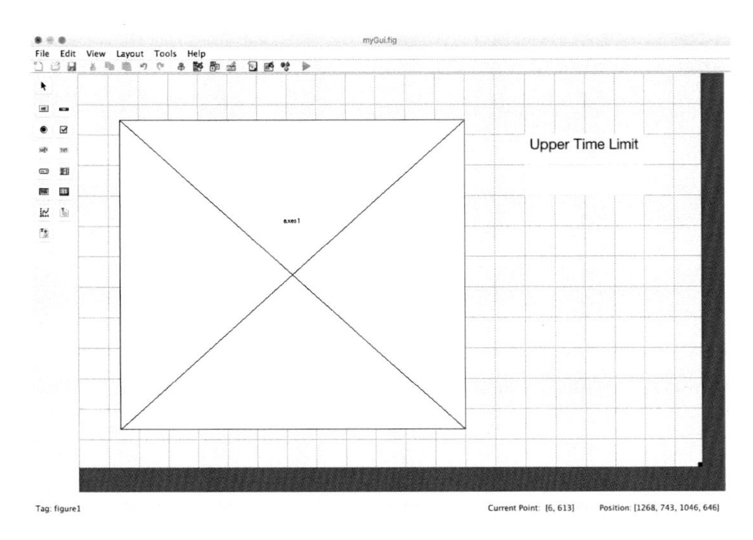

FIGURE 11.8 **The `guide` editor showing the GUI canvas with added static and edit text components.**

The first and second callbacks relate to the creation and deletion of the component and the third is called when a mouse click takes place over the component.

The edit text box has two further callback functions with the following names:

- `KeyPressFcn`
- `Callback`

The first of these is called when the user presses *any* key inside the editable text box.

The second callback function (with the unimaginative name '`Callback`' !) is the more useful one; it is called when the user has entered some text into the box *and* subsequently hits the *Enter* (or *Return*) key. The *Enter* key signifies that the user has finished typing and wants to pass the typed information to the program. Next we will use this function to obtain user input.

■ Activity 11.9

O11.B, O11.C

Getting user input with a `Callback` function: Hit *Save* on the `guide` and highlight the edit text box. Right-click or go to *View*, then choose *View Callbacks* → `Callback` so that the `guide` automatically generates the starting code for this function.

The code for the `Callback` function of our edit text box should be automatically generated and added to the *myGUI.m* file. It contains the definition line and some comments on usage. These should include hints on how to use the function:

```
function edit1_Callback(hObject, eventdata, handles)
% hObject     handle to edit1 (see GCBO)
% eventdata   reserved - to be defined in a future version of MATLAB
% handles     structure with handles and user data (see GUIDATA)

% Hints: get(hObject,'String') returns contents of edit1 as text
%        str2double(get(hObject,'String')) returns contents of ...
     edit1 as a double
```

■

The function name for the specific text box is `edit1_Callback`.[2] The usage hints give an idea of how to access what the user enters into it. As before, the first argument to the function is a handle to our text box, which in this case is called *edit1*.

As described in the help text, we can access what the user enters as a standard text string using:

```
get(hObject, 'String')
```

This uses the `get` function to ask for a property named 'String' that the text box object has. See Section 11.3 for a description of the `get` (and `set`) function.

If, on the other hand, we want to interpret what the user has entered as a number we can repeat the above and pass what `get` returns to the built-in `str2double` function. This converts the string to a double precision floating point number:

```
str2double(get(hObject,'String'))
```

If the user enters a string that cannot be interpreted as a number (e.g. 'xQSd!') then `str2double` will return NaN (see Section 5.3).

The snippet of code that we will place into the callback for the edit text box to allow the user to modify the plot is shown below. In this code, the variable `tUpper` represents the user-defined upper limit for the time axis value to plot:

```
%...
tUpper = str2double(get(hObject,'String'));

if ( isnan(tUpper) || tUpper <= 0.01 )
  return
end
% ...
```

[2] If we add a second text box, the function would be called `edit2_Callback` and so on.

This code retrieves the user input as a string, converts it to a numeric value and assigns it to `tUpper`. If the string was not interpretable as a number or is too small, the function will simply return. Otherwise, the code goes on to modify the plot as shown below.

■ Activity 11.10

Place the following code in the `edit1_Callback` function so that the user can modify the plotted time range.

O11.B, O11.C

```
function edit1_Callback(hObject, eventdata, handles)
% ... usage text here ...

% START ADDED CODE
tUpper = str2double(get(hObject,'String'));

if ( isnan(tUpper) || tUpper <= 0.01 )
  return
end

% Read data
ecg = csvread('example_ecg_data_b.txt');
nPts = length(ecg);

% Sampling frequency
sampFreq = 720;   tStep = 1/sampFreq;

% Time points at given sampling frequency.
t = 0:tStep:(nPts-1)*tStep;

% Plot data:
plot(handles.axes1, t, ecg);

xlim([0, tUpper]);  %%% Set time range's upper limit!

% END ADDED CODE
```

■

This code takes the user input string and tries to convert it to a number. If this is not possible (i.e. if `tUpper` is NaN) then the function simply returns. Otherwise, the function re-plots the data and sets the limit on the horizontal axis to go from zero to the user's value of `tUpper` (using the `xlim` function).

Note how we have used the third argument to the callback function (the one named `handles`) to access the axes component on the GUI so that we can plot into it.

Run the code above and experiment by changing the upper time limit in the edit text box. The box should start off empty but you should be able to insert a number into the box and hit *Enter*. The plot should be updated to reflect the value you entered.

Inserting a non-numeric value into the box, e.g. 'blah', should not affect the plot. The same should be true if you insert negative values or any small values less than 0.01.

11.2.3 Maintaining State and Avoiding Duplicated Code

In Activity 11.5 we added the first code for plotting to the axes (in the function `myGui_OpeningFcn`).

In Activity 11.10, we added code to the function `edit1_Callback` to update the plot if the user sets an upper time limit in the edit text box.

This means that we now have *two* places in the code that draw a plot on the axes. It would be good to avoid such unnecessary duplication of code so we will now make a single plot function and call it from the other parts of the code when it is needed.

In order to do this, we need to be able to pass information between the different parts of the code. In particular, we will need to keep a record of the state of variables that belong to the GUI and we would like this state to persist between successive calls to the callback functions. We will do this in the next example by using the `handles` variable.

■ Activity 11.11

O11.C

Add a general plotting function: Start by adding a function to the end of the script *myGui.m*. Even though this is a new function, the code inside the function's body should be fairly familiar by now!

```
function myPlot(handles)
% Plots the data with current parameters to the axes.
% Read data
ecg = csvread('example_ecg_data_b.txt');
nPts = length(ecg);

% Sampling frequency
sampFreq = 720; tStep = 1/sampFreq;

% Time points at given sampling frequency.
t = 0:tStep:(nPts-1)*tStep;

% Plot data:
plot(handles.axes1, t, ecg);

xlim([0, handles.tUpper]);
```

Note how this function expects to receive a handles structure and that it also expects that one of the fields of the structure is called `handles.tUpper` that contains the upper time limit to use in the plot. The set of axes to place the plot in is represented by the field `handles.axes1` as before. ■

Now that we have made a general plotting function, we need to remove the statements that plot the data from the original code and replace them with calls to our new function `myPlot`. We need to call our new plotting function from two places in the code:

- Inside the opening function `myGui_OpeningFcn` (Activity 11.12).
- Inside the edit text box callback function `edit1_Callback` (Activity 11.13).

■ Activity 11.12

O11.B, O11.C

Modify the GUI opening callback: In the first place, remove the code that plots inside `myGui_OpeningFcn`, i.e. *delete* the code that was added in Example 11.5. This should be straightforward as this code is between the lines `% START ADDED CODE` and `% END ADDED CODE`. Then replace it so that the body of the function looks like the following:

```
% Choose default command line output for myGui
handles.output = hObject;

% START ADDED CODE
handles.tUpper = 5.0;
myPlot(handles)
% END ADDED CODE

% Update handles structure
guidata(hObject, handles);
```

■

In the next task, we will modify the code for the text box callback function.

■ Activity 11.13

O11.B, O11.C

Modify the function `edit1_Callback`: For the `edit1_Callback`, we need to delete the code that was added in Activity 11.10. Again this should be easy as it is between the lines `% START ADDED CODE` and `% END ADDED ... CODE`. Then edit it so that it now reads as follows:

```
% START ADDED CODE
tUpper = str2double(get(hObject,'String'));

if ( isnan(tUpper) || tUpper <= 0.01 )
  return
end

handles.tUpper = tUpper;
myPlot(handles);

% Update handles structure - Important!
guidata(hObject, handles);
% END ADDED CODE
```

■

In this code, we still read in the user input and assign the local variable `tUpper` as before. This time, however, we also assign the value to a field in the `handles` structure that has the same name.[3] We can then call our plotting function which will make use of the updated value of `handles.tUpper`. Note also that, this time, we have included a call to a function called `guidata` – this call was already automatically included in `myGui_OpeningFcn` so we did not need to add it there (see Activity 11.12). In the current callback function, we need to explicitly add this call. Check that the code is running properly by testing it with a range of upper time limits, including non-numeric values.

The `guidata` function is important as it ensures that the updates we make to the variable `handles` *persist* for the GUI as a whole. This means that when the current callback function, or indeed any other callback function is invoked later, the updated values of the `handles` fields remain accessible.

If we add more code later that also adds information as fields in the `handles` variable, then we will also need to make a call to the `guidata` function at the end, using the same pair of arguments (`hObject` and `handles`).

11.2.4 Tidying Up

In the next activity, we tidy up by ensuring the content of the edit text box is set on loading the GUI. We will revisit the figure and *m*-files for this GUI in some of the exercises at the end of the chapter.

■ Activity 11.14

O11.B, O11.C

Ensuring the edit text box shows the correct value on starting the GUI: We saw in Activity 11.7 how we can set what is displayed in a text box using the property inspector. Here we will control what is shown in the edit text box in code so that, as the GUI is being created and loaded, it displays the initial upper limit.

To do this, add one line to the body of the GUI's opening function, `myGui_OpeningFcn`.

```
% START ADDED CODE
handles.tUpper = 5.0;
myPlot(handles)

set(handles.edit1, 'String', num2str(handles.tUpper))
% END ADDED CODE
```

The extra line makes a call to the `set` command in order to adjust the edit text box's 'String' property. See Section 11.3 for a description of the use of the `set` function. Here, we simply put in the starting value for the upper time limit. Test that the code is working properly, i.e. that the correct value is displayed when the GUI starts. ■

---

[3] It does not need to be the same name, but it is descriptive so there is no problem with using it twice, once for the local variable and once for the field name in `handles`.

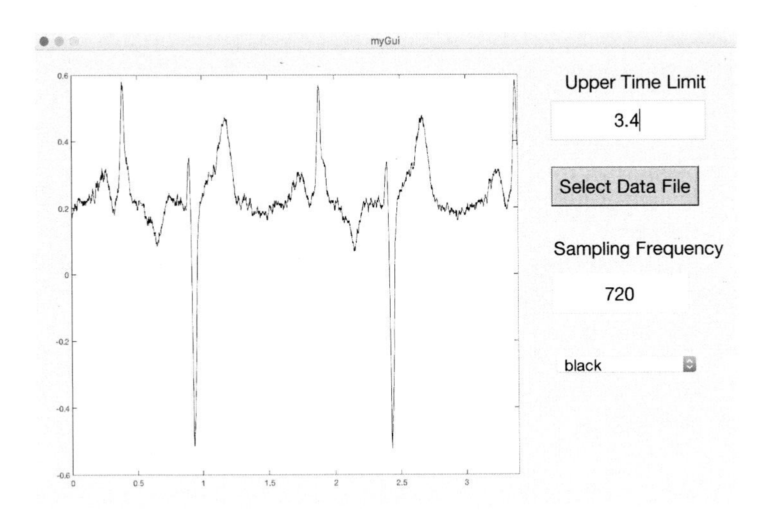

FIGURE 11.9 An illustration of the GUI that has been used in the activities in this chapter. It also shows further components that will be added in the exercises at the end of the chapter.

So far, we have created a GUI and added a set of axes to it for plotting and a pair of text boxes to control the upper time limit used when plotting the ECG data. In the exercises, we add further components to add more options for user interaction. An illustration of the final GUI after all components have been included, from the activities above and the later exercises, is given in Fig. 11.9.

11.3 HANDLES

MATLAB and a number of other languages use *handles* to access a variety of different objects. In the commands below, we create a handle to the line object that is created when we plot some data.

```
x = 1:10; y = sqrt(x);
h = plot(x,y, '-o')
```

In the above, the `plot` command is used. In contrast to how we have used `plot` in the past, this time we collect a return argument and assign it to `h`.

The variable `h` is a *handle* to the line in the plot. We can find out about how it represents the line's information with the `get` function. If we type `get(h)` in the command window, a long list of properties belonging to the line is printed to the screen.

We can ask for specific properties with the `get` command as well. For example, we can ask which marker is currently being used and what the line width is:

```
>> get(h, 'Marker')
ans =
o
>> get(h, 'LineWidth')
ans =
    0.5000
```

We can also modify the line through its handle by using the set command. For example, typing the following into the command window will directly modify the style of the plotted line:

```
set(h, 'Marker', 'x'),    set(h, 'MarkerSize', 10)
set(h, 'LineStyle', ':'), set(h, 'LineWidth', 2)
```

In the illustration above, we have looked at a handle for a line object. Handles can also be used with other objects, such as figures, axes, text boxes etc. Whichever kind of object is represented, we can use the get and set commands to access and modify the properties of the specific object that a handle refers to.

11.4 SUMMARY

MATLAB provides a set of tools for creating Graphical User Interfaces (GUIs). We have seen how we may use the guide tool to design a GUI by positioning and sizing components such as text boxes and axes. Further components such as buttons and pop-up menus can also be added.

In general, when we add a component, we need to know how to adjust the way it behaves and interacts with the user. This is done using *callback functions*. It is possible to find out which callback functions are available by reading the documentation but it is generally easier to find out using the context menu by right-clicking on the component that we have added.

The context menu will show which callbacks are available. Selecting one of them from the context menu will cause MATLAB to automatically generate code for the callback along with hints in the usage text. Reading this gives an idea of how to use the component.

The automatically generated skeleton code for the callback functions, plus any code that we add, will control the behavior of the GUI. Each call is invoked when specific *events* occur, such as the user clicking inside a plot or entering text into an edit text box.

Another way to find out how to add code to a callback function is to place break points in its body. This allows us to inspect the contents of the variables passed in as arguments (such as handles) while the code is running. We can experiment with modifying the code or variables when the debugger is stopped in the callback's function body.

11.5 FURTHER RESOURCES

- MathWorks File Exchange: http://www.mathworks.co.uk/matlabcentral/fileexchange.
- The MATLAB documentation has some examples of GUI building under *GUI Building → GUI Building Basics → Examples and Howto*. The online version can be found at: http://uk.mathworks.com/help/matlab/gui-building-basics.html. It would be a useful exercise to work through these as well as the examples in this chapter.

EXERCISES

■ Exercise 11.1

O11.A, O11.B, O11.C

Create a new blank GUI using the `guide` tool and carry out the following steps:

1. Add two text boxes to the GUI: one static text box and one edit text box.
2. Save the GUI in a folder called guiEx1.
3. Run the GUI to make sure it works.
4. Close the GUI and change the code in the edit text box callback function so that the string entered by the user is displayed in the other (static) text box.

Hints:

- *The* `handles` *variable in the automatically generated code will have fields for each of the edit and static text boxes. These should be named* edit1 *and* text1 *(although the number might vary).*
- *Use the* `set` *and* `get` *functions to set and access the* `'String'` *properties in the text boxes.* ■

■ Exercise 11.2

O11.A, O11.B, O11.C

Use the `guide` to create a blank GUI. Save it as *guiEx2*, i.e. the `guide` will create files *guiEx2.m* and *guiEx2.fig*. Add two edit text boxes (their tags should be *edit1* and *edit2*) and a static text box (its tag should be *text1*).

1. Write code so that the edit text boxes show a (string) value of 1 at opening.
2. Write code so that the static text box shows the sum of the two numbers shown in the edit text boxes.
   Hints:
   - *Use the* `handles` *structure to store numeric values for the contents of the edit text boxes.*

- *Use a separate function called* `runOperation` *to actually carry out the adding of the numbers and the updating of the static text box.*

3. Write code to guard against non-numeric input in the edit text boxes, i.e. ensure that all stored and displayed values are still numeric even if the user makes a mistake. ■

■ Exercise 11.3

O11.A, O11.B, O11.C

Continuing with the GUI created in Exercise 11.2, use the `guide` to add a Button Group to the GUI. Add four Radio Buttons to it. We will use each button for one of the four basic arithmetic operations, so modify them so that their strings are `'add'`, `'multiply'`, `'subtract'`, and `'divide'`. Modify the string of the panel containing the buttons so that it says 'operation'.

Add code to the script and modify the `runOperation` function so that the correct operation is applied to the current values in the edit text boxes.

Hints:

- *Add a field to* `handles` *to store the current operation.*
- *Use a* `switch` *statement in* `runOperation` *to select which operation to apply.* ■

■ Exercise 11.4

O11.C

This exercise continues from the previous work on the *myGui* plotting GUI (Activity 11.1 to Activity 11.14).

Modify the code in *myGui.m* so that, if a user gives a non-numeric value in an edit text box, the value displayed is re-set to the correct number. That is, the edit text box for the upper time limit should show the current value of the upper time limit (instead of the non-numeric value that the user entered). ■

■ Exercise 11.5

O11.A, O11.B, O11.C

This exercise continues the previous exercise on the *myGui* plotting GUI (Activity 11.1 to Activity 11.14).

In this exercise, you will add a push button to the GUI that, when clicked, will allow the user to select a file of their choice containing the data to plot. Any file containing a single column of numbers can be loaded and plotted. You can use the file *example_ecg_data_c.txt* which is available on the web site. Download it and save it to a folder, perhaps even a different folder from the one containing the *myGui* code.

Open the `guide` tool and load `myGui.fig` in the design view. You may need to browse to the directory where it was saved. Use the `guide` to place a push button component onto the GUI.

Using the property inspector, set the text for the push button to *Select Data File* and ensure the background color is set to gray. (The background color may remain white when editing but should turn gray on running the GUI.)

Even though we have not yet added functionality to the push button, save the GUI figure and test that it runs. The save step should generate some automatic code for the button. It is likely to be called `pushbutton1_Callback` but the number may vary.

Add code to the push button callback function that allows the user to select a text file. Begin by using the `uigetfile` command as follows:

```
[filename,pathname] = uigetfile({'*.txt'});
```

This opens a file browser for the user and, if the user chooses a text file, returns the file name and the path to the file in the two output arguments. If the user does not select a file, then the name will equal zero. You can use this information to include a check on the user input. Use the output from `uigetfile` to construct the full file name (see the `fullfile` command).

You will need to maintain state by storing the name of the data file being used. In order to do this, save the file name in the GUI's `handles` data structure. See Activity 11.13 for an example of how to maintain state.

The only other part of the *myGui.m* code that currently uses the data file is the plotting function `myPlot`. It currently uses a hard-coded string to refer to the original data file (*example_ecg_data_b.txt*). Change its code so that it uses the user set data file.

In the callback function for the push button, after saving the data file in the `handles` structure, ensure that a call to the plotting function is made so that the display is updated.

Finally, to make the code start up the GUI correctly, the data file stored in the `handles` data structure needs to be initialized. Do this in the general GUI opening callback function (near where the upper time limit is initialized and before the first call to plot). ■

■ Exercise 11.6

O11.A, O11.B, O11.C

This exercise is a continuation of the GUI exercises and examples on the *myGui* interface (Activity 11.1 to Activity 11.14).

Recall that the default sampling frequency of the ECG data was hard-coded in the file *myGui.m*. It is set to a value of 720 in the main opening callback. In earlier tasks, we added a choice of input data file. This means that it might be useful to allow the user to change the sampling frequency of the data.

Add another pair of text boxes to the GUI using the `guide` tool: one static text box and one edit text box. Use the edit text box to take a number that is used for the sampling frequency of the function plotted (see the similar work for the upper time limit in Activity 11.13). The number will need to be stored in the `handles` structure in a suitable field. Initialize the sampling frequency stored in `handles` in the GUI opening function to the default value (720).

Ensure the stored value of the sampling frequency is used in the plotting function instead of the hard-coded value.

Set the string for the edit text boxes in the script during the call to the opening function for the GUI (see Activity 11.14). ■

■ Exercise 11.7

O11.A, O11.B, O11.C

This exercise continues from the previous exercises and also uses the *myGui* plotting GUI.

Insert a pop-up menu onto the GUI to control the color of the plot. The choice should be between black, red, blue or green.

Hints:

- *Use the* `handles` *object to store a field for the plot color. This should be the single letter string that MATLAB uses for plot colors:* `'k'`, `'r'`, `'b'`, `'g'`.
- *Initialize the field in the* `handles` *object in the GUI opening function to a default color letter, say* `'k'`.
- *Set the* `'String'` *property for the pop-up menu in the main GUI opening function. Assign a cell array of strings to contain what is displayed to the user, for example* `{'black', 'red', 'blue', 'green'}`.
- *The hints for the pop-up menu callback function should give an idea of how to retrieve which string the user has chosen.*
- *Update the* `myPlot` *function so that it uses the plot color field from the* `handles` *structure when it actually calls the* `plot` *function.* ■

FAMOUS COMPUTER PROGRAMMER: LINUS TORVALDS

Linus Torvalds is a Finnish computer scientist, born in 1969 in Helsinki, Finland. He is famous for being the developer of the LINUX open-source operating system. As a child Torvalds was always interested in home computers. He owned several and enjoyed writing software for them including games and a text editor. He went on to study Computer Science at the University of Helsinki.

In 1991, whilst still a student at Helsinki, Torvalds purchased a personal computer (PC), which ran the Microsoft DOS operating system. But Linus preferred the UNIX operating system (see Chapter 7's Famous Computer Programmer) that he had used on the university's computers. Therefore, he set himself the task of writing a version of UNIX for PCs. After he had produced the core of the new operating system, he made the source code freely available on the internet, and asked other programmers around the world to contribute to the project. Many took up the offer and the LINUX project was born. Torvalds initially wanted to call his open-source operating system FREAX (free UNIX), but his friend Ari Lemmke who administered the FTP server where the software was hosted for download didn't like this name and so called it LINUX (Linus' UNIX). These days, LINUX is one of the most widely used operating systems in the world. Versions of LINUX are run by many PCs, servers and smart phones.

The fact that the source code of the LINUX kernel was released freely under the GNU software license (see Chapter 9's Famous Computer Programmer) meant that Torvalds never benefited financially from the success of LINUX. However, in 1999, two leading commercial developers of LINUX-based software, Red Hat and VA Linux, presented Torvalds with stock options in gratitude for his creation, which made him a wealthy man.

Torvalds currently lives in Portland, Oregon, USA, and works full-time for the LINUX Foundation on improving LINUX.

> *"Most good programmers do programming not because they expect to get paid or get adulation by the public, but because it is fun to program."*
>
> **Linus Torvalds**

CHAPTER 12

Statistics

LEARNING OBJECTIVES

At the end of this chapter you should be able to:

- *O12.A* Use MATLAB to compute numerical summary statistics of univariate and bivariate data
- *O12.B* Use MATLAB to display visualizations of univariate and bivariate data
- *O12.C* Use MATLAB to carry out appropriate hypothesis tests on statistical data and interpret their results

12.1 INTRODUCTION

When writing programs to process biomedical data it is often desirable to be able to summarize and draw conclusions from the data. This might be, for example, to report numerical summary statistics for a set of measurements, or to identify associations between different variables. Of course, it is always possible to export any data we have produced with MATLAB (see Chapter 6) and then to analyze it with our favorite dedicated statistical package. However, this is not usually necessary, as the MATLAB Statistics and Machine Learning Toolbox covers most of the more commonly used statistical techniques. Therefore, in this chapter we give an overview of the statistical capabilities of MATLAB.

Note that, because this book is primarily about computer programming, we will not be discussing statistical theory in detail: for this we refer the interested reader to one of the excellent textbooks on the theory and application of statistics (see Further Resources at the end of this chapter). We focus instead on the practical side of using statistical techniques in MATLAB and give a number of biomedical examples for illustration.

12.2 DESCRIPTIVE STATISTICS

We start off with *descriptive statistics*, meaning any technique (numerical or visual) that can be used to describe a set of measurements we have made of one

MATLAB Programming for Biomedical Engineers and Scientists. DOI: 10.1016/B978-0-12-812203-7.00012-4

Table 12.1 MATLAB functions for computing summary statistics of central tendency and dispersion

Measures of central tendency		Measures of dispersion	
`mean(x)`	Mean value of array	`std(x)`	Standard deviation of array values
`median(x)`	Median value of array	`var(x)`	Variance of array values
`mode(x)`	Mode value of array	`iqr(x)`	Inter-quartile range of array values
		`range(x)`	Range of array values

or more variables. We refer to such a set of measurements as a *sample*, and these samples are typically drawn from a larger *population* of such values. Note the distinction here with *inferential statistics* (see Section 12.3), in which we try to *infer*, or draw conclusions about the population based on the sample.

We will separately cover descriptive statistics of *univariate* numerical data (i.e. in which we have one variable per measurement) and *bivariate* data (i.e. two variables per measurement). For each of these we discuss different numerical tools for summarizing the data and techniques for visualizing the data.

12.2.1 Univariate Data

For univariate data there are two types of numerical statistic that we can use to summarize our measurements: measures of *central tendency* and measures of *dispersion*. Measures of central tendency summarize the average value of the sample, whereas measures of dispersion summarize the spread of the values around this average. Table 12.1 lists the different MATLAB functions for computing these measures.

■ Example 12.1

O12.A

The following code illustrates the computation of various measures of central tendency and dispersion from body temperature data (in Celsius) that have previously been saved to a *MAT* file.

```
% load body temperature data
load('temperatures.mat'); % loads in 'temps' variable

% compute measures of central tendency
fprintf('Mean temperature = %.1f\n', mean(temps));
fprintf('Median temperature = %.1f\n', median(temps));
fprintf('Mode temperature = %.1f\n', mode(temps));

% compute measures of dispersion
fprintf('Std dev of temperature = %.1f\n', std(temps));
fprintf('Variance of temperature = %.1f\n', var(temps));
fprintf('IQR of temperature = %.1f\n', iqr(temps));
fprintf('Range of temperature = %.1f\n', range(temps));
```

■

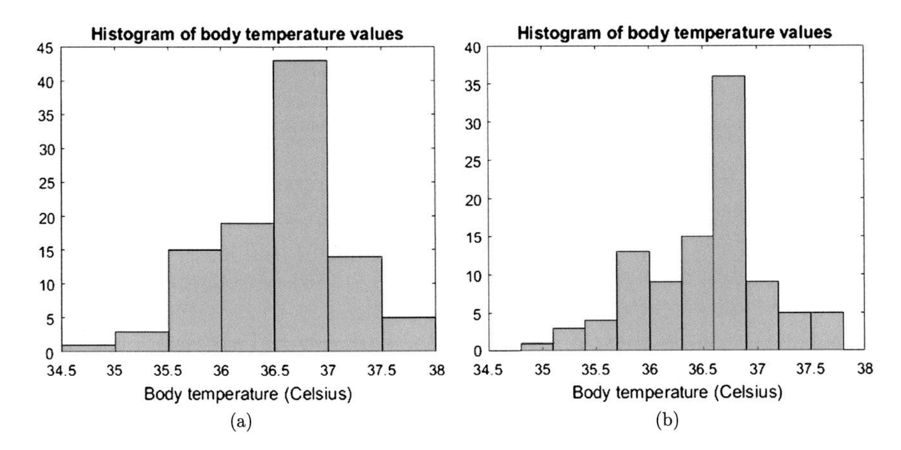

FIGURE 12.1 Histograms of body temperature: (a) using 5 bins; (b) using 10 bins.

We have already reviewed MATLAB visualization approaches in Chapters 1 and 8. In this chapter, we introduce some further visualizations that can be useful when analyzing data as part of a statistical analysis. For univariate data the most common way to visualize multiple measurements of a variable is the *histogram*. Histograms allow us to visualize the distribution of values in the measurements.

■ Example 12.2

Histograms can be displayed for any variable which takes discrete or continuous values. In MATLAB we use the `histogram` function, as illustrated below.

O12.B

```
% load body temperature data
load('temperatures.mat'); % loads in 'temps' variable

% dispay histograms
close all;
figure;
% binning algorithm chooses 5 bins for these data
histogram(temps);
title('Histogram of body temperature values');
xlabel('Body temperature (Celsius)');
figure;
% Explicity set the number of bins to 10
histogram(temps, 10);
title('Histogram of body temperature values');
xlabel('Body temperature (Celsius)');
```

Fig. 12.1 shows the output of this code. As can be seen, histograms can appear quite different depending on the number of bins used. The `histogram` function uses an algorithm to automatically estimate a good number of bins to use with the data. In the second call, we take control of this and set the number of bins to 10 (using the optional second argument). ■

12.2.2 Bivariate Data

For bivariate data that are continuous, discrete or ordinal, we can measure the *correlation* between the two variables. Commonly applied measures of correlation are the Pearson's correlation coefficient (for continuous or discrete data that are normally distributed) and Spearman's (rank) correlation coefficient (for bivariate data in which at least one variable is not normally distributed or is ordinal). Both of these measures can be computed in MATLAB using the `corr` function, as illustrated in the following example.

■ Example 12.3

O12.A

Research has suggested that body temperature can reduce with age. Furthermore, studies of dementia patients indicate a relationship between body temperature and the decline in cognitive function. The following code investigates these possible links using data loaded in from *MAT* files. (Note that these data were not measured from real patients but were artificially generated based on values reported in the literature.)

Cognitive function is commonly measured using the Mini Mental State Examination (MMSE), a 30-point questionnaire that tests a range of cognitive tasks. The data used in this example represent age, body temperature and MMSE scores for 100 dementia patients. As the histograms shown in Figs. 12.1a and 12.2 illustrate, the temperature and age variables appear to be normally distributed, whereas the MMSE data appear skewed. Therefore, to measure the correlation between body temperature and age we can use Pearson's correlation coefficient, but to measure the correlation between body temperature and MMSE score we should use Spearman's correlation coefficient. Consider the following code.

```
% load data
load('temperatures.mat');
load('ages.mat');
load('mmse.mat');

% compute Pearson's correlation
[rhoA, pvalA] = corr(ages,temps)

% compute Spearman's correlation
[rhoB, pvalB]=corr(temps,mmse, 'type', 'Spearman')
```

Note that, by default, the `corr` function computes Pearson's correlation coefficient. However, if we specify an extra pair of *name–value* arguments (`'type', 'Spearman'`) it will compute the Spearman's correlation instead.

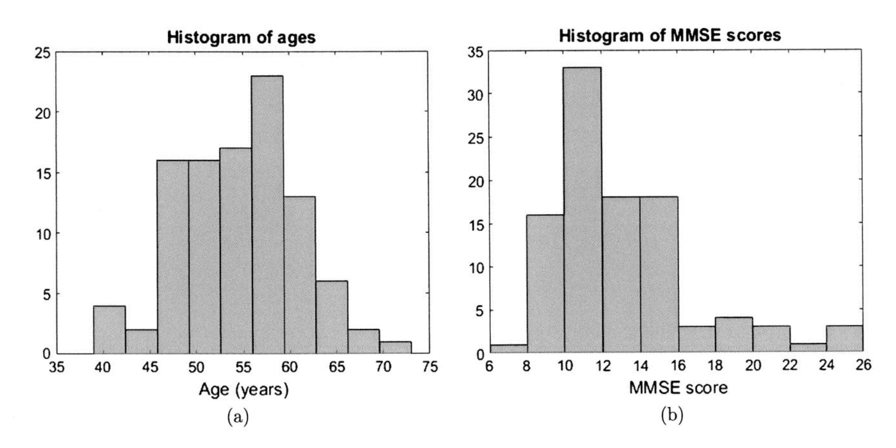

FIGURE 12.2 Histograms of (a) age; (b) MMSE score.

The `corr` function returns two values: `rho` represents the correlation coefficient, and `pval` represents the *'p-value'* of the correlation. This represents the *significance*, i.e. the probability that the correlation observed is due to random variation in the data. A low p-value indicates that the correlation found is likely to represent a real association between the variables in the underlying population.

You can download this code and data from the book's web site. The data have a negative correlation (−0.3409) between body temperature and age, with a low p-value of 5.1832e−04. The correlation between body temperature and MMSE score is 0.1149, with a p-value of 0.2551. Therefore, based upon these results, we can conclude that there is likely to be a real (but weak) negative correlation between body temperature and age. We cannot assume any correlation between body temperature and MMSE score as there is a 25% chance that we would have seen this pattern of variation in the data even if no correlation between the two variables existed. ■

There are a number of ways to visualize univariate and bivariate data. For bivariate data, perhaps the obvious choice, when the data consist of *corresponding* (or *paired*) observations, is simply to plot one variable against the other for all observations in the sample. This is known as a *scatter plot* and an example is shown in Fig. 12.3a. The scatter plot allows us to visually confirm the negative correlation between body temperature and age that we found in Example 12.3.

If we wish to compare the measures of central tendency and dispersion of two symmetrically distributed discrete or continuous variables, we can use an *error bar* plot. An example of this is shown in Fig. 12.3b, which illustrates that the dementia patients who were aged sixty or less had a slightly higher body temperature than those who were over sixty. The crosses represent the mean values and the error bars represent the standard deviations.

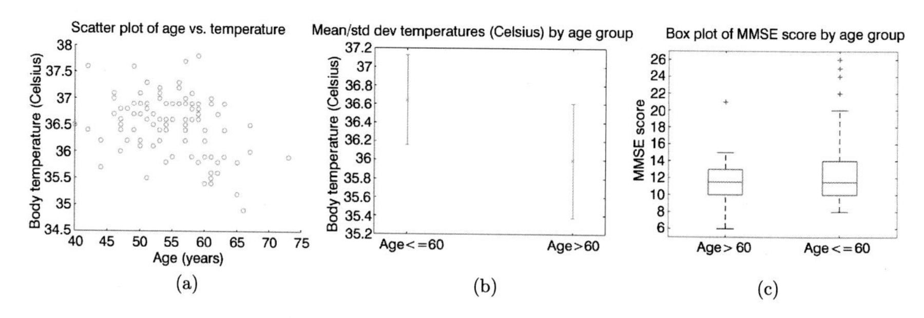

FIGURE 12.3 Visualizations of multivariate data: (a) scatter plot of age against body temperature; (b) error bar plot of body temperature for different age groups; (c) box plot of MMSE score for different age groups.

Finally, to compare measures of central tendency and dispersion across sets of skewed data, we can use a box plot, which visualizes the medians and inter-quartile ranges of two (or more) variables. An example of this is shown in Fig. 12.3c. The red lines represent the medians and the blue boxes the inter-quartile ranges. The black lines indicate an estimate of the extent of the distribution while the red crosses indicate possible outlier values.

■ Example 12.4

O12.B

This example illustrates how MATLAB can be used to generate the visualizations shown in Fig. 12.3. The `scatter` function is used to create a scatter plot and takes two array variables of the same length (paired data). The `errorbar` function is used to create an error bar plot. Note the use of logical indexing (see Section 9.3.3) when producing this plot. This divides the temperature data into two arrays: one for patients aged sixty or less and one for patients over sixty. The `boxplot` function produces a box plot, and takes two arguments: the array of values that are to appear on the y-axis (MMSE scores in our case) and a second array of the same length that will be used to group the values in the first array. We used the same logical array of age classifications for this purpose.

```
% load data
load('temperatures.mat');
load('ages.mat');
load('mmse.mat');

% scatter plot
close all;
figure;
scatter(ages, temps);
xlabel('Age (years)');
ylabel('Body temperature (Celsius)');
title('Scatter plot of age vs. temperature');
```

```
% error bar plot
age_class = ages<=60; % logical array
means = [mean(temps(age_class)) mean(temps(~age_class))];
stdevs = [std(temps(age_class)) std(temps(~age_class))];
figure;
errorbar(means, stdevs, 'x');
ylabel('Body temperature (Celsius)');
set(gca, 'XTick', 1:2);
set(gca, 'XTickLabel', {'Age<=60','Age>60'})
title('Mean/std dev temperatures (Celsius) by age group');

% box plot
figure;
boxplot(mmse, age_class);
ylabel('MMSE score');
set(gca, 'XTick', 1:2);
set(gca, 'XTickLabel', {'Age>60','Age<=60'})
title('Box plot of MMSE score by age group');
```

■

■ Activity 12.1

O12.A, O12.B

The enlargement of brain ventricles has been proposed as an indicator of Alzheimer's disease progression. The change in volume of the brain ventricles can be measured from 3-D medical images (e.g. MR and/or CT scans). The file *Alzheimers_data.mat* contains the result of a study in which 17 Alzheimer's patients had their disease *progression* measured using two methods: the change in MMSE score and the change in ventricle size. Each patient was measured at two time points, and the file contains the changes in MMSE score (first column, positive values show symptoms becoming worse) and the changes in ventricle volume (second column, in cm^3). Analyze these data as follows:

1. Produce histograms of the two variables, i.e. change in MMSE score and change in volume. What do the histograms suggest?
2. Visualize the data using a scatter plot. What does the scatter plot show?
3. Calculate an appropriate measure of correlation between the two variables. Also determine the p-value of the correlation. ■

12.3 INFERENTIAL STATISTICS

When working with biomedical data, as well as visualizing and summarizing, we often want to try to answer a specific question using the data we have as evidence. For example, based on data we might collect on the response to a new treatment, can we say that the treatment works better than an existing treatment or a placebo? This type of question brings us into the realm of *inferential statistics*, in which we try to *infer* something about the population of interest based on a sample we have drawn. In this section we review some of the functions that MATLAB provides to help us answer such questions, which are also known as *hypothesis tests*. Hypothesis tests help us to decide whether a given hypothesis can be rejected (or not) given a set measurements we have in our data. Typically, the hypothesis that we actually test is known as the *null hypothesis*, and the alternative to it is known as the *alternative hypothesis*.

12.3.1 Testing the Distributions of Data Samples

Before we look at specific questions such as whether a treatment works or not, it is important to know something about the *distribution* of our data. We have already seen how histograms can be used to visually inspect data distributions (see Section 12.2.1), but in inferential statistics we want to be more precise than this. In order to be able to choose an appropriate hypothesis test for answering questions using the data, it is first important to be able to test the hypothesis that the data sample was drawn from a population with a particular distribution. The MATLAB function `chi2gof` is one way to do this. It performs a *chi square* (χ^2) *goodness-of-fit test* on the data. This tests the null hypothesis that the data sample comes from a particular distribution, for a given level of confidence (e.g. 95%). The alternative hypothesis is that the data does not come from such a distribution. A common distribution to test for in inferential statistics is the *normal*, or *Gaussian* distribution, as the following example illustrates.

■ Example 12.5

O12.C

An experiment is performed to test the efficacy of a new dietary supplement intended to lower cholesterol levels in humans. Total cholesterol level data in mmol/L was gathered from 125 subjects who suffered from high cholesterol. The readings were obtained before subjects took the new supplement and three months afterward. The data are available in the files *cholesterol_pre.txt* and *cholesterol_post.txt*. We want to test if the distributions of the two variables (i.e. the total cholesterol levels before and after) are normally distributed. Examine the following code.

```
% Null Hypothesis = data drawn from a normal distribution
%  with same mean and variance as sample
% Alternative Hypothesis = data does not come from such a
%  distribution

% load data
chol_pre = load('cholesterol_pre.txt');
chol_post = load('cholesterol_post.txt');

% always a good idea to visualise data first ...
subplot(2,1,1);
histogram(chol_pre);
xlim([2 9]);
```

```
title('Histogram of cholesterol levels before treatment');
subplot(2,1,2);
histogram(chol_post);
xlim([2 9]);
title('Histogram of cholesterol levels after treatment');

% Apply chi square goodness-of-fit tests
% with default 95% confidence
%   h==1 Reject null hypothesis
%   h==0 Do not reject
[h_pre, p_pre] = chi2gof(chol_pre)
[h_post, p_post] = chi2gof(chol_post)
```

Note that the `chi2gof` function returns two arguments: `h` refers to the result of the hypothesis test (1 means reject the null hypothesis, 0 means do not reject), whereas `p` is the p-value of the test, i.e. the probability of seeing this level of variation in the data if the null hypothesis were true. The default confidence level to work to is 95%, but we can change this by using optional *name–value* pairs of arguments as shown below (an `alpha` value of 0.01 corresponds to 99% confidence).

```
[h_pre, p_pre] = chi2gof(chol_pre, 'alpha', 0.01)
```

In this example, for both variables the return value `h` was 0 meaning that the null hypothesis could not be rejected. Therefore, we assume for the purposes of our statistical analysis that both variables are normally distributed. ■

See the Further Resources section at the end of this chapter for details of alternative ways of using MATLAB to test if our data are normally distributed.

12.3.2 Comparing Data Samples

The main reason for testing whether a data sample was drawn from a population with a normal distribution is to help us to decide which type of test to apply for further questions based on the data. Two categories for such tests are *parametric* and *nonparametric* tests. Broadly speaking, we can apply parametric tests if all of our data are normally distributed, and we must apply nonparametric tests otherwise.

There are three very common tests for comparing data samples, depending on the number of samples we have and their relationship to each other:

- A one sample t-test against an expected value;
- A two sample *paired* t-test, in which there exist one-to-one correspondences between values in the first sample and values in the second sample;
- A two sample *unpaired* t-test, in which no such correspondences exist.

Each of these tests can be either *one-tailed* or *two-tailed*. With a one-tailed test we are trying to decide whether the values in the population from which one sample was drawn are *larger* or *smaller* than either a specific value or the set of values in another sample. With a two-tailed test we are looking for *any* difference regardless of whether it is larger or smaller.

The most commonly applied parametric hypothesis test is the Student's t-test. MATLAB functions for applying a Student's t-test for the three different situations are summarized below.

- `[h,p] = ttest(x,m,'alpha',α,'tail',t)`: Performs a one sample t-test between the dataset `x` and the expected value `m`. The null hypothesis is that there is no difference between the mean of `x`, and `m`. The alternative hypothesis is that there is such a difference. An `h` value of 1 is returned if the null hypothesis can be rejected, and an `h` of 0 means it cannot be rejected. The variable `p` contains the p-value, which represents the probability that the null hypothesis is true and that the sample data occurred by chance. The extra optional pairs of arguments, `'alpha',α`, and `'tail',t`, specify the significance level and type of the test. The default value for the significance level α is 0.05, representing 95% confidence. The default value for the type `t` is `'both'`, meaning a two-tailed test. If `t` is `'right'` it will perform a one tailed test for the mean of `x` being greater than `m`. It `t` is `'left'` it will perform a one-tailed test for the mean of `x` being less than `m`.
- `[h,p] = ttest(x,y,'alpha',α,'tail',t)`: Performs a two sample *paired* data t-test between the datasets `x` and `y`. The null hypothesis is that the mean difference between pairs of `x` and `y` values is zero. The alternative hypothesis is that this mean is not zero. An `h` value of 1 is returned if the null hypothesis can be rejected and an `h` of 0 means it cannot be rejected. The extra optional argument pairs specifying α and `t` have similar meanings to those described above. A `'tail'` of `'left'` tests if the mean of `x` is less than the mean of `y`, whilst `'right'` tests if the mean of `x` is greater than the mean of `y`.
- `[h,p] = ttest2(x,y'alpha',α,'tail',t)`: Performs a two sample *unpaired* data t-test between the datasets `x` and `y`. The null hypothesis is that `x` and `y` come from normal distributions with the same mean. The alternative hypothesis is that their means are different. An `h` value of 1 is returned if the null hypothesis can be rejected. The extra optional argument pairs specifying α and `t` have the same meanings as described above.

Note that we use the same function (`ttest`) to perform both a one sample t-test against an expected mean, and a two sample paired data t-test. The difference lies in the nature of the second argument supplied: if it is a scalar value a one-sample t-test will be performed, whereas if it is an array of the same length as the first argument a two-sample paired data t-test will be performed.

■ Example 12.6

In this example we extend the cholesterol case study we introduced in Example 12.5. Now that we are confident that the data are normally distributed, we want to use parametric tests to answer the following questions:

O12.C

- Is the total cholesterol level *before* taking the supplement significantly different to the 'healthy' level of 5 mmol/L?
- Is the total cholesterol level *after* taking the supplement significantly different to the 'healthy' level of 5 mmol/L?
- Is the total cholesterol level after taking the supplement significantly lower than the total cholesterol level before taking the supplement?

We will work to a 95% confidence level. The following code illustrates how to answer these questions using MATLAB.

```
% load data
chol_pre = load('cholesterol_pre.txt');
chol_post = load('cholesterol_post.txt');

% Test both datasets against expected value of 5mmol/L
% 2-tailed, 95% confidence
% Null Hypothesis = there is no difference between the
%  sample mean and 5
% Alternative Hypothesis = there is such a difference
% (h=1 means reject null hypothesis, h=0 means do not reject)
[h_pre5, p_pre5] = ttest(chol_pre, 5)
[h_post5, p_post5] = ttest(chol_post, 5)

% Test for decrease in cholesterol after taking supplement
% 1-tailed, 95% confidence
% Null Hypothesis = mean difference between pre and post
%  data not < 0
% Alternative Hypothesis = mean difference between pre
%  and post data < 0
% (h=1 means reject null hypothesis, h=0 means do not reject)
[h_cmp, p_cmp] = ttest(chol_pre, chol_post, 'tail', 'right')
```

Note that for the first two questions we perform a two-tailed test, as we are looking for *any* difference, but for the third question we are looking for one variable to be less than the other so we perform a one-tailed test. We use a paired data t-test as the pre-supplement and post-supplement cholesterol levels were measured from the same subjects. In this example we are able to reject the null hypothesis for all three questions, meaning that both variables are different to 5 mmol/L and that the post-supplement cholesterol level is lower than the pre-supplement level. ■

If at least one of our data samples was drawn from a population which is not normally distributed we cannot use a Student's t-test. In this case a non-parametric test should be used. MATLAB also provides functions for common nonparametric tests that broadly correspond to the three types of parametric test we introduced above. These are summarized below:

- `[p,h]= signrank(x,m,'alpha',α,'tail',t)`: Performs a one sample Wilcoxon signed rank test for a single sample `x` against a specified population median `m`. The null hypothesis is that there is no difference between the median of `x`, and `m`. The alternative hypothesis is that there is such a difference. An `h` value of 1 is returned if the null hypothesis can be rejected. The extra optional argument pairs specifying α and `t` have similar meanings to those described earlier for the `ttest` and `ttest2` functions. If the value of `t` is `'both'` (the default value) the test will be two-tailed. If `t` is `'right'` it will perform a one tailed test for the sample median being greater than `m`. If `t` is `'left'` it will perform a one-tailed test for the sample median being less than `m`.[1]
- `[p,h]= signrank(x,y,'alpha',α,'tail',t)`: Performs a two sample paired data Wilcoxon signed rank test between the datasets `x` and `y`. The null hypothesis is that the median of the differences between `x` and `y` is zero. The alternative hypothesis is that there is a non-zero difference. An `h` value of 1 is returned if the null hypothesis can be rejected. The extra optional argument pairs specifying α and `t` have similar meanings to those described earlier: a `'tail'` of `'left'` tests if the median of `x` is less than the median of `y`, whilst `'right'` tests if the median of `x` is greater than the median of `y`.
- `[p,h] = ranksum(x,y,'alpha',α,'tail',t)`: Performs a two-tailed unpaired Mann–Whitney U test between the datasets `x` and `y`. The null hypothesis is that `x` and `y` have the same distributions and medians. The alternative hypothesis is that their medians are different but the distributions are otherwise identical. An `h` value of 1 is returned if the null hypothesis can be rejected. The extra optional argument pairs specifying α and `t` have the same meanings as they do for the `signrank` function.

In the above, note the reversed ordering of the output arguments `p` and `h` when compared to the `ttest` and `ttest2` functions.

■ Example 12.7

O12.C

To illustrate the use of nonparametric hypothesis tests we introduce a new case study. A company develops a new surgical technique for inserting hip implants. The angular accuracy of the placement can be assessed postoperatively using X-ray imaging. Accuracy data (in degrees) have been gathered from a cohort of 100 patients. Half of the cohort (randomly selected) underwent the new surgery and the other half underwent traditional surgery. The data are available in the file *implant_data.mat*. This contains two MATLAB variables: `trad_errors` and `new_errors`, which represent the angular

---

[1] *N.B.* The one sample Wilcoxon signed rank test assumes symmetrically distributed differences. If this is not the case, a sign test can be performed with the function `signtest`.

errors for the traditional surgery group and the new technique's group respectively.

Assuming that at least one of the two samples was drawn from a population that is not normally distributed, we want to perform an appropriate hypothesis test to determine, with 95% confidence, if the new surgical technique results in lower errors than the traditional technique. Examine the following code.

```
% load data
load('implant_data.mat');

% Test for decrease in error with new technique
% Unpaired data, so Mann Whitney U test
% 1-tailed, 95% confidence
% Null Hypothesis = median of new errors not less than
%  median of trad errors
% Alternative Hypothesis = median of new errors less than
%  median of trad errors
% (h=1 means reject null hypothesis, h=0 means do not reject)
[p,h] = ranksum(trad_errors, new_errors, 'tail', 'right')
```

We use a Mann–Whitney U test. Even though the two samples we are comparing have the same length (50), there is no pairing between the patients who underwent traditional surgery and those who underwent the new surgical technique. Therefore, we have *unpaired* data. We are looking for a *decrease* in errors so we apply a one-tailed test. For this data, `ranksum` returns an `h` value of 1, meaning that we reject the null hypothesis and conclude that the new technique has lower errors. ■

■ Activity 12.2

O12.B, O12.C

Bilirubin is a substance found in the blood which is the breakdown product of the clearance of aged red blood cells. A study is investigating a possible link between high bilirubin levels and the risk of developing cardiovascular disease.

A group of subjects were randomly selected 10 years ago and data have been gathered from them over the past 10 years. For each subject, the data consists of the average serum bilirubin level over the 10 year period (in mg/dL) and whether or not the subject developed a cardiovascular disease during the 10 year period (1 = yes, 0 = no). The data are available in the file *bilirubin_data.mat*, and consists of two variables: `bilirubin` and `cardiovascular_disease`.

Write code to split the bilirubin level data into two groups: one for those subjects who had cardiovascular disease, and one for those who didn't. Visualize the distributions of the data values in these two groups. What are your comments?

Perform an appropriate hypothesis test to determine, with 95% confidence, if the bilirubin levels of those who had cardiovascular disease are from a different distribution to the bilirubin levels of those who didn't. Explain why you chose the test you did and comment on the result. ■

12.4 SUMMARY

MATLAB provides functions for most common statistical tasks, either through its core functionality or through the Statistics and Machine Learning Toolbox.

Table 12.1 summarizes the functions available to numerically summarize either the central tendency or the dispersion of a univariate data sample. The correlation between two variables can be measured using the `corr` function.

The distribution of a univariate sample can be visualized using a histogram of its values, which in MATLAB can be produced using the `histogram` function. Multivariate data visualizations can be produced using the `scatter`, `errorbar` and `boxplot` functions.

As well as numerically summarizing and visualizing data (descriptive statistics), MATLAB also provides functions for answering questions about the population from which the data were drawn (inferential statistics). A chi square goodness-of-fit test can be used to test if a univariate sample fits a specific distribution such as the normal distribution. The MATLAB function `chi2gof` can be used for this purpose. The results of such a test help us to decide whether subsequent tests should be parametric or nonparametric.

It is possible to carry out parametric hypothesis tests using MATLAB. The Student's t-test can be performed using the `ttest` and `ttest2` functions. The `ttest` function is used for one sample comparisons against an expected mean value, and also for two sample comparisons using paired data. The `ttest2` function is used for two sample comparisons using unpaired data.

For nonparametric tests, the MATLAB functions `signrank` and `ranksum` perform, respectively, a Wilcoxon signed rank test (either one sample or two sample paired data) and a Mann–Whitney U test (two sample unpaired data).

All hypothesis tests can be applied using different confidence levels and as either one-tailed or two-tailed tests.

12.5 FURTHER RESOURCES

- "Vital Statistics: An Introduction to Health Science Statistics" by Stephen McKenzie [6] is an excellent and accessible introduction to the theory and application of statistics with a focus on the health sciences.

- We introduced `chi2gof` as one way of checking to see if our data were drawn from a normally distributed population. The following are alternatives:
  - Kolmogorov–Smirnov test: see the MATLAB documentation for `kstest`.
  - Lilliefors test: see the MATLAB documentation for `lillietest`.
  - Z-test: see the MATLAB documentation for `ztest`.
  - Shapiro–Wilk test: there is no built-in MATLAB function but implementations are available from the MathWorks File Exchange (http://uk.mathworks.com/matlabcentral/fileexchange).
- See the MATLAB documentation for more details on the functions we have covered, and for other useful functions:
  - Descriptive statistics: http://uk.mathworks.com/help/stats/descriptive-statistics.html.
  - Statistical visualization: http://uk.mathworks.com/help/stats/statistical-visualization.html.
  - Hypothesis testing: http://uk.mathworks.com/help/stats/hypothesis-tests-1.html.

EXERCISES

■ Exercise 12.1

O12.A, O12.B, O12.C

A new drug has been developed to suppress the desire for drinking alcohol among alcoholic patients. Data have been gathered of the number of units of alcohol consumed weekly by a group of 73 alcoholics before and after treatment with the drug. The data are in the file *alcohol.mat*, and consists of two variables: `units_before` and `units_after`.

In this exercise, you will use MATLAB to determine if the alcohol intake figures are significantly lower after treatment compared to before treatment. This will involve analyzing the data to determine if they are normally distributed, numerically summarizing the data and then choosing and applying an appropriate hypothesis test.

1. Load in the data and visualize the distributions of the alcohol consumption data, both before and after taking the drug.
2. Perform appropriate hypothesis tests to determine if the before treatment and after treatment data are normally distributed. Work to a 95% confidence level.
3. Choose and compute appropriate numerical summary statistics of the before and after treatment data. Justify your choices.
4. Choose an appropriate hypothesis test and apply it to determine if the after treatment data are significantly lower than the before treatment data. Work to a 95% confidence level. Explain your choice of hypothesis test and comment on the result. ■

■ Exercise 12.2

O12.A, O12.C

A large scale study of the UK population has gathered data from volunteers through questionnaires, a variety of tests and imaging studies. A research team would like to use these data to investigate a potential link between diet and cardiac health.

Data gathered from 5000 subjects are available to you in the file *heart.mat*. The file contains two variables. The `diet` variable is a classification of each subject's diet into one of three categories: good (= 2), average (= 1) or poor (= 0). The `ef` variable is a measurement of each subject's *ejection fraction* as measured using dynamic magnetic resonance imaging. Ejection fraction represents a measure of how much blood is pumped out of the heart in each beat, and is considered to be a good indicator of cardiac health. Typical (healthy) values are between 50% and 65%, but people with poor cardiac health are likely to have lower values.

Use MATLAB to perform the following tasks.

1. Load the data and visualize them using four histograms included in the same figure. These should show respectively:
   - All ejection fraction values;
   - Ejection fraction values for subjects with a poor diet;
   - Ejection fraction values for subjects with an average diet;
   - Ejection fraction values for subjects with a good diet.

   Comment on the distributions that the histograms show.
2. Compute an appropriate measure of correlation between diet and ejection fraction. Also, compute the probability that this correlation is due to random variation in the data. Comment on the results.
3. Perform appropriate hypothesis tests to determine, with 95% confidence, if the ejection fraction data for subjects with poor, average and good diets were drawn from normal distributions.
4. Based on the results of these three tests, choose and apply appropriate hypothesis tests to determine if:
   - the ejection fractions of subjects with poor diets are significantly different to the ejection fractions of subjects with good diets;
   - the ejection fractions of subjects with poor diets are significantly less than 50%.

   In the above, as well as calling the appropriate MATLAB functions, interpret the outputs to give your conclusions in plain English. ■

■ Exercise 12.3

O12.A, O12.B, O12.C

Positron emission tomography (PET) imaging enables in-vivo quantitative measurements of radiotracer concentration, and is commonly applied to monitor the progress of cancer. A common measure is the standardized uptake value (SUV). If the SUV of a tumor increases over time this means the cancer is becoming more active.

1. SUV values can be affected by the presence of motion artefacts in the PET images, so correcting for motion such as that caused by breathing is of interest. A research group has developed a new technique for respiratory motion correction of PET images. PET images have been acquired from 100 patients and SUV values calculated for their tumors. These data are in the file *suv_pre_correction.txt*. The images were then motion corrected using the new algorithm and the SUV values recalculated. The recalculated values are in the file *suv_post_correction.txt*. Using MATLAB, read in the data, visualize it and choose and apply an appropriate hypothesis test to determine if the new technique changes the SUV values with 95% confidence.
2. SUV values for clinical use are sometimes computed by either taking the mean or maximum SUV value within a region of interest. These approaches may be subject to noise in the images. A better way may be to use a statistical test based on all values within the region. A patient undergoing chemotherapy has had PET images acquired before and after treatment. SUV values for all voxels within a tumor region for the two images can be found in the files *suv_pre_chemo.txt* and *suv_post_chemo.txt*. Because the tumor may have changed size the two files may contain different numbers of SUV values and the values do not represent corresponding voxels. Doctors wish to know if the activity of the tumor has changed as a result of the treatment. First, use MATLAB to compute mean and maximum SUV values for before and after chemotherapy to answer this question. Next, perform an appropriate statistical hypothesis test to answer the same question with 95% confidence. ■

FAMOUS COMPUTER PROGRAMMERS: LARRY PAGE AND SERGEY BRIN

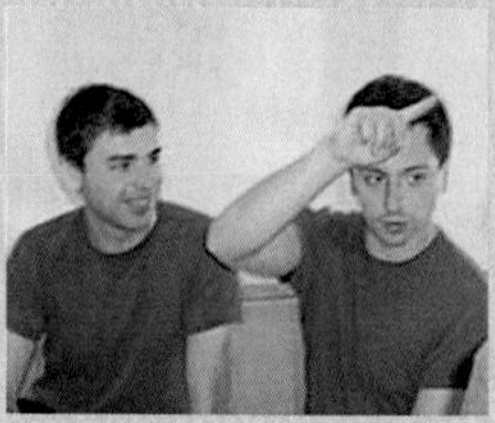

For this final chapter we have two famous computer programmers: Larry Page and Sergey Brin. Page and Brin are the co-founders of Google. Through founding Google and developing much of the core software upon which it is based, these two American computer scientists have revolutionised the way in which we use the internet.

Initially Google was solely an internet search engine, although it has now come to include a wide range of other functionality, such as email and social networking. The Google search engine started off as a research project in 1996 whilst Page and Brin were both PhD students at Stanford University. Its main innovation was a technology known as PageRank, which used much more sophisticated ways of determining a web page's relevance to a search term. Most search engines at that time used relatively simplistic algorithms that just counted how many times a search term appeared in the page. PageRank also incorporated information about the network of hyperlinks between relevant pages.

Page and Brin did most of the programming for PageRank in their dormitory room which was crammed with computers. The project soon began to place a big load on Stanford's computing infrastructure, but by this time Page and Brin had decided to end their studies and focus solely on the search engine. They scraped together funds from friends and family, bought some servers and rented a garage to host them. Shortly after this they founded the Google company and gained further investment. Today Google has an estimated value of over US$100 billion.

"Solving big problems is easier than solving little problems."

Sergey Brin

References

[1] M. Studenski, Effective dose to patients and staff when using a mobile PET/SPECT system, J. Appl. Clin. Med. Phys. 14 (3) (2013) 215–225.

[2] C. Le Tourneau, J.J. Lee, L.L. Siu, Dose escalation methods in phase I cancer clinical trials, J. Natl. Cancer Inst. 101 (10) (2009) 708–720.

[3] Y. Nakamoto, M. Osman, C. Cohade, L. Marshall, J. Links, S. Kohlmyer, R. Wahl, PET/CT: comparison of quantitative tracer uptake between germanium and CT transmission attenuation-corrected images, J. Nucl. Med. 43 (9) (2002) 1137–1143.

[4] D. Persaud, J.P. O'Leary, Fibonacci Series, Golden Proportions, and the Human Biology, HWCOM Faculty Paper 27, 2015.

[5] C.J. Mode (Ed.), Applications of Monte Carlo Methods in Biology, Medicine and Other Fields of Science, InTech, 2011.

[6] S. McKenzie, Vital Statistics: An Introduction to Health Science Statistics, 1st edition, Churchill Livingstone, 2013.

Index

Symbol

A

B

C

D

E